Process Engineering
in Biotechnology

The Biotechnology Series

This series is designed to give undergraduates, graduates and practising scientists access to the many related disciplines in this fast developing area. It provides understanding both of the basic principles and of the industrial applications of biotechnology. By covering individual subjects in separate volumes a thorough and straightforward introduction to each field is provided for people of differing backgrounds.

Titles in the Series

Biotechnology: The Biological Principles: M.D. Trevan, S. Boffey, K.H. Goulding and P. Stanbury
Fermentation Kinetics and Modelling: C.G. Sinclair and B. Kristiansen (Ed. J.D. Bu'Lock)
Enzyme Technology: P. Gacesa and J. Hubble
Animal Cell Technology: M. Butler
Fermentation Biotechnology: O.P. Ward
Genetic Transformation in Plants: R. Walden
Plant Biotechnology in Agriculture: K. Lindsey and M.G.K. Jones
Biosensors: E. Hall
Biotechnology of Biomass Conversion: M. Wayman and S. Parekh
Process Engineering in Biotechnology: A.T. Jackson

Upcoming Titles

Biotechnology Safety
Plant Cell and Tissue Culture

Applied Gene Technology
Bioreactors

Series Editors

Professor J.A. Bryant   *Department of Biology, Exeter University, England*
Professor J.F. Kennedy   *Department of Chemistry, Birmingham University, England*

Series Advisers

Professor C.H. Self   *Department of Clinical Biochemistry, University of Newcastle upon Tyne, England*

Dr R.N. Greenshields   *G.B. Biotechnology, Swansea, Wales*

The Institute of Biology I⦾B

This series has been editorially approved by the **Institute of Biology** in London. The Institute is the professional body representing biologists in the UK. It sets standards, promotes education and training, conducts examinations, organizes local and national meetings, and publishes the journals **Biologist** and **Journal of Biological Education**.

For details about Institute membership write to: Institute of Biology, 20 Queensberry Place, London SW7 2DZ.

Process Engineering in Biotechnology

A.T. Jackson

Open University Press
Milton Keynes

Open University Press
22 Ballmoor
Celtic Court
Buckingham MK18 1XW

First Published 1990

British Library Cataloguing in Publication Data

Jackson, A. T.
    Principles of biotechnological engineering.
    1. Chemical engineering. Biochemical reactions
    I. Title   II. series
    600.63

ISBN 0–335–15811–0
ISBN 0–335–15810–2 (pbk) O 95867 ·

Typeset by Vision Typesetting, Manchester
Printed in Great Britain by Biddles Limited, Guildford and King's Lynn

Contents

Preface

The design of large-scale biotechnological processes involves many different disciplines and requires the integration and interaction of a multitude of different parts of the sciences, technologies, and engineering if the design is to be brought to commercial fruition.

One important aspect of this multidisciplinary collaboration is communication between the participants; familiarity with and understanding of the terms used, together with an appreciation of the potential problems of each discipline, helps to lubricate the collaborative process, as well as saving time.

This book has been written with the objective of introducing non-engineers (mainly biologists) to basic chemical engineering (process engineering) principles. It is not intended to 'convert' other disciplines to engineering but to act as a guide to some of the engineering principles involved in the translation of laboratory-generated data and relationships into the reality of an operating full-scale plant.

I have tried to keep the mathematics as simple as possible so that a basic knowledge of simple integration and differentiation is the only mathematical prerequisite.

If, as a result of reading this book, I have saved someone from asking an irrelevant question, or, more importantly, helped someone to ask a more relevant question, then I will feel that my efforts will have been amply rewarded.

My thanks are due to Stockdale Filtration Systems Ltd of Macclesfield for supplying and allowing me to reproduce Figs 6.3, 6.4, 6.5, and 6.6, and to Westfalia Separator Ltd of Milton Keynes for providing Figs 6.9 and 6.10.

I also gratefully acknowledge the assistance of my colleagues, in particular Dr M.W. Austin and the late Mr F.H. Cass for their encouragement and suggestions, and to the large number of students over the past eight or nine years who unknowingly assisted in developing the treatment I have adopted.

Last, but by no means least, to my wife for her fortitude in becoming a 'writing' widow, and to Mr Alan Sugar whose vision of computers-for-all made the task of writing somewhat easier than it might have been.

A.T. Jackson

Chapter 1

Introduction

Industrial biotechnology is ultimately concerned with using biological materials to generate useful products. The development of a new process begins with the identification of a need, for example of a new antibiotic, food additive, source of protein, etc. The next step is the development of the biological knowhow, for example the screening of a group of micro-organisms for the new product, genetic manipulation, and physiological studies. All of these endeavours are, however, without point, unless there exists the technology to utilize the biology in an industrial process; this is the stage where chemical engineering comes in.

Chemical engineering is the engineering of processes in which materials undergo a change. This change may be chemical or physical, and the task of the chemical engineer is to take scientific ideas developed in the laboratory and to turn them into *economic* realities.

At the heart of a biotechnology production process is the 'bioreactor' where the biotransformation takes place. Biotransformations may be classified into a number of different types:

- the production of biomass for direct use; for example, yeast for brewing or baking
- the use of an organism to synthesize *de novo* a useful material from simple nutrients; for example, the production of an antibiotic, enzyme, or alcohol by cells growing in a medium (substrate) containing sugars, amino acids, trace elements
- the use of whole, sometimes viable, cells to effect complex transformations on specific precursors added to the medium; for example, the conversion of digitoxin to digoxin by cells of *Digitalis*
- simple chemical conversions using non-viable cells or enzymes; for example, the isomerization of glucose to (sweeter) fructose by the enzyme *xyloseisomerase*.

Whatever the type of biotransformation there will be a need for both 'upstream' processes (media preparation and sterilization, reactor inoculation) and 'downstream' processes (for separation, purification, and packaging of the product). In practice, the total cost of a process is governed more by the costs of downstream processing than by the cost of the bioreactor part of the operation.

Feasibility studies of new processes must always include an *economic balance* of the money, materials, and manpower used against the value of the products obtained. Included in every design must be the optimization of energy requirements and the optimal control of all parts of the process.

The total cost of a product produced by a biological process is made up of the following cost contributions:

- cost of raw materials (nutrients, additives, control chemicals, packaging)
- capital costs of equipment (fermentors, upstream and downstream process equipment, waste disposal, instrumentation and control)
- operating costs (labour, repairs and maintenance, overheads)
- services costs (cooling water, refrigeration, electricity, steam, chilled water, sterile air, inert gas)
- research and development costs.

There are engineering problems associated with all stages of the scaling-up of biological processes from a laboratory scale to an industrial scale. The large size of industrial fermentors required for commercial viability (up to $200 \, m^3$) gives rise to problems of mixing, heat transfer, and control, as well as the problems of physically handling large volumes of liquids and gases. The concentration of products obtained from biological processes is low (typically 10 per cent maximum), and recovery processes often require a large energy input. The optimization and integration of all parts of the process, from the initial fermentation to marketable product, is essential.

There are a wide range of materials produced from microbiological sources ranging from antibiotics and food to chemicals. Some processes are aerobic and require a supply of air to large volumes of liquid, quite frequently under aseptic conditions.

The fermentation stage exhibits similarities in all cases (whether aerobic or anaerobic), and many of the upstream and downstream processing operations are similar for a wide range of different processes, only quantities and components differ.

The approach which is used by chemical engineers in the design of processing equipment is the concept of *unit operations*. Unit operations were conceived in the late 1880s and have been developed in such a way that the same basic principles can be applied to the same operation irrespective of the material being processed or of the overall process. The only factors which change from process to process are physical properties and flowrates. In general, the physical properties are found to affect the efficiency of the process, the flowrates and quantities affecting the physical size of the equipment.

Some examples of unit operations are:

- heat transfer
- fluid mixing
- separation methods
  - physical—filtration, centrifugation
  - diffusional—evaporation, extraction, adsorption, distillation, drying
- pumping of liquids.

The intention of this book is to give an appreciation of the unit operations involved in biological processing, and the physical, chemical, and biological properties of the materials which are of significance to the engineer to allow the satisfactory design of an economically optimum plant for the large-scale production of useful products.

Most of the information required for design can be generated in the laboratory, quite often during the experimental development of the biological knowhow. This information can then be applied using the well-established unit operations based on the present chemical and biochemical industries, together with some newer, more specific biotechnological operations to economically scale-up the processes developed in the laboratory.

The rest of this chapter is intended to give some idea of the range of processes in use today for both the large-scale manufacture of chemicals (including pharmaceuticals) as well as for the manufacture of high-quality speciality products (enzymes). Details of the microbiology of the processes are outside the scope of this text.

Industrial processes utilize either a single, specially selected species of micro-organism, or a specially selected mixed microbial population, and the following sections have been classified on this basis.

Single Species Processes

ENZYME MANUFACTURE[1, 2]

A large part of the biochemical industry is devoted to the manufacture of synthetic enzymes, i.e. the deliberate growth of a micro-organism under conditions which produce the desired enzyme, rather than the extraction of naturally occurring enzymes.

β-Amylase

Specially selected and cultivated strains of *Aspergillus niger* are grown in an aerobic, deep fermentation process to produce extracellular β-amylase. Since the enzyme appears in the liquid component of the fermentation broth (mycelial suspension), the downstream processing, i.e. the recovery of the enzyme, concentrates on the treatment of the liquid phase, and a typical process diagram can be built up out of the following operations:

- the fermentation stage
- separation of the mycelium from the liquid
- concentration of the liquid enzyme solution
- production of solid product (for example, precipitation onto starch)
- drying of the solid product.

This type of process is shown schematically in Fig. 1.1.

In general, the scheme shown in Fig. 1.1 is followed for most enzyme production processes. Differences occur depending on the biology of the system and the form of final product required, and the products mentioned in the following sections have been chosen to illustrate these differences.

α-*Amylase*

Specially selected strains of *Bacillus subtilis* are grown in an aerobic deep fermentation process to produce the extracellular enzyme. The product is usually marketed as a concentrated, standardized solution, and the recovery of the product follows a similar pattern to that shown in Fig. 1.1, but stops after the concentration stage.

A change in substrate nutrient balance and a different aeration pattern during fermentation causes the micro-organism to synthesize proteolytic enzymes (proteases) which are marketed in both liquid and solid forms for use in household detergents (usually as a solid) or for the dehairing of hides prior to leather production.

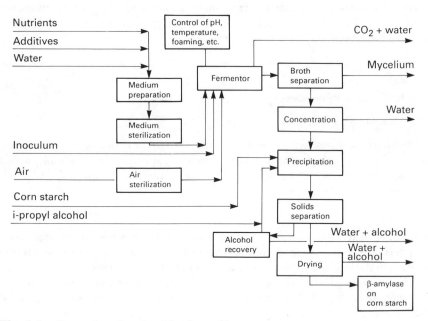

Fig. 1.1 Operations involved in β-amylase manufacture.

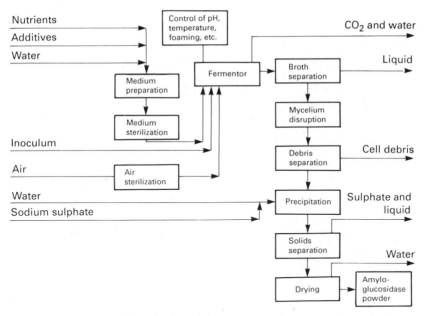

Fig. 1.2   Production of amyloglucosidase.

Amyloglucosidase
A. niger grown in an aerobic deep fermentation process produces this enzyme, but intracellularly. Although the fermentation stage is similar to that used for the production of β-amylase, the recovery of the enzyme requires extra processing of the mycelial mass after initial broth separation.

The mycelium is macerated to rupture the cells and to release the enzyme, and a further separation stage is required to separate the cell debris. The enzyme in the liquid stream from this second separation is then precipitated (normally using ionic salts), and the precipitate separated and dried to give a highly active enzyme powder. This type of process is shown diagrammatically in Fig. 1.2.

ANTIBIOTICS MANUFACTURE

There are about 50 antibiotic compounds produced commercially using biological methods. Most are produced using moulds (e.g. penicillin) although a few are produced using bacteria (e.g. streptomycin).

One of the problems in the 'working-up' (downstream processing) of antibiotics is that they are extremely sensitive to heat, and thermal concentration methods (which can be used for some thermally stable enzymes) are unsuitable for antibiotics. In these cases, the use of liquid extraction processes using refrigerated organic solvents for the concentration of the antibiotic ensures minimum thermal damage to the product. For example, penicillin is extracted into amyl or butyl acetate and crystallized at temperatures around 0 °C.

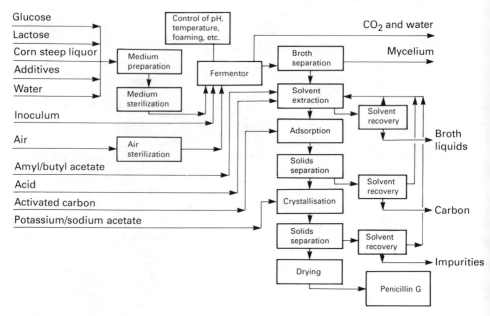

Fig. 1.3   Penicillin manufacture.

One of the advantages of using liquid extraction is that if a careful selection of solvent is made, the impurities (byproducts) can be selectively removed from the system, thus ensuring a high-purity product.

Figure 1.3 shows the stages involved in the recovery of penicillin.

BULK CHEMICALS

Acetone/Butanol[3]

These two solvents (together with a trace of ethanol) can be produced from either molasses (sugar waste) or corn starch liquor using *Clostridium* species in an anaerobic fermentation. Prior to the growth of the petrochemical industry, this process formed the major source of these solvents. The fermentation itself is susceptible to phage contamination, and the maintenance of strict sterile conditions is essential. After fermentation, the solvents are recovered using a multi-column, fractional distillation train, the first distillation combining the processes of cell removal and initial solvent concentration. A schematic diagram of this process is shown in Fig. 1.4.

Ethyl Alcohol[3]

The production of potable ethyl alcohol is carried out in an anaerobic process using *Saccharomyces* strains with a sugar substrate (normally molasses). An alternative substrate is starch (from grain) and prior to fermentation the starch is enzymically converted to fermentable sugars.

As with the acetone/butanol process, the alcohol is recovered by fractional distillation, producing purified alcohol/water solutions ranging from 55 per cent to 98 per cent (by weight) depending on the end use.

Acetic Acid[4]

Using *Acetobacter* species, ethyl alcohol in solution (8–12 per cent) can be converted aerobically to acetic acid solution for use in the food industry as 'vinegar'.

If the alcohol used to produce the vinegar has been manufactured biologically, then the product is referred to as 'malt vinegar'.

The production of acetic acid in tonnage quantities for use in the chemical industry is more economically carried out chemically by the oxidation of acetaldehyde.

Citric Acid[3]

The production of citric acid using *A. niger* was originally carried out using static aerobic surface culture in shallow trays. In recent years, a two-stage deep fermentation process has been developed.

The recovery process involves the separation of mycelium from the broth containing the citric acid, followed by the addition of a small controlled quantity of lime to precipitate the oxalic acid byproduct (at 80–90 °C). After separation of the calcium oxalate, the solution is further treated with lime (at 95 °C) to precipitate calcium citrate.

The separated salt is acidified with sulphuric acid, the liquid then being concentrated and citric acid crystallized out. A diagram of this multi-precipitation process is shown in Fig. 1.5.

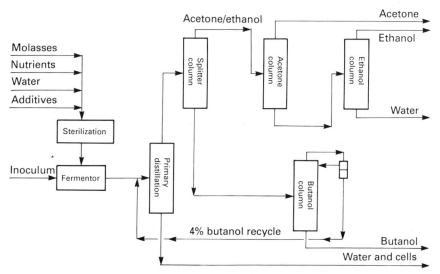

Fig. 1.4 Production of acetone/butanol by fermentation.

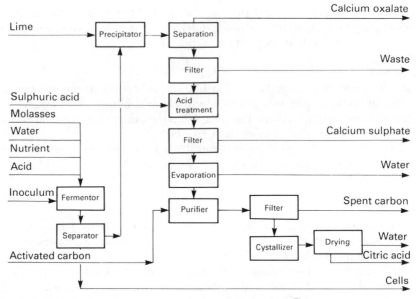

Fig. 1.5 Operations involved in citric acid production.

Lactic Acid[5]

This is produced by anaerobic fermentation using *Lactobacilli* species. It is an interesting process because the microbial reaction must be carried out at a pH of about 7 (neutral). Because the lactic acid produced reduces the pH, the fermentation is carried out in the presence of neutralizing calcium ions (Ca^{2+}), and any acid produced is immediately converted to the calcium salt. There are no competing organisms for the substrate, and the process is not carried out aseptically. The high cost of lactic acid is due to the recovery and conversion of the calcium salt to the acid and the removal of byproducts. The downstream processing involves not only separation, concentration, etc., but more sophisticated techniques such as ion exchange (both anionic and cationic). The final part of the process uses solvent extraction with i-propyl alcohol which is a selective solvent for the impurities. Re-extraction of the i-propyl alcohol solution with water yields a purified 80–85 per cent lactic acid solution.

YEAST PRODUCTION

Baker's Yeast[6]

This is produced using *Saccharomyces cerevisiae* in an aerobic fermentation at aeration rates designed to give an oxygen concentration in the substrate greater than 0.2 ppm. Aerobic operation encourages the production of yeast cells and minimizes the production of alcohol. The preferred substrate is made from molasses with added mineral supplements. The supplements are chosen to give a cell product which can be readily compressed into blocks for ease of transport and storage.

Single-cell Protein (SCP)[7, 8]

The development of processes designed to provide biomass (cells) as an alternative source of protein for both human and animal nutrition has involved the investigation of a number of micro-organism/substrate combinations. The growth of *Torula* and *Candida* yeasts on sugars and hydrocarbons was investigated in the 1970s, and the 1980s has seen the development of sophisticated fermentation processes designed to utilize alcohols and other petrochemical industry products as the main nutrient in the substrate. A single-cell protein process is described in more detail in Chapter 9.

BREWING AND WINEMAKING[9, 10]

Wine production is a mixed aerobic/anaerobic process using either wild yeast strains present on the grape skins or selected yeast cultures for inoculation. The yeast is first grown under aerobic conditions to stimulate the growth of the yeast cells and is then changed to anaerobic operation to stimulate the production of alcohol and carbon dioxide.

Beer production is similar to winemaking (mixed aerobic/anaerobic) but a large part of the processing is involved with the production of a suitable substrate for the yeast inoculum (upstream processing).

The grain (barley) is allowed to germinate for about three days, encouraging the formation of amylolytic and proteolytic enzymes, and it is then carefully dried for storage unless the malting process is immediately followed by the mashing process (substrate preparation).

The substrate for the yeast (wort) is prepared by carefully warming the malted grain with water (often with added starches) so that the action of the enzymes produces a mixture of simple sugars, amino acids, and flavour compounds. The wort is then boiled which has a concentrating and sterilizing effect. The wort may be boiled with hop flowers to give a distinctive flavour (and to incorporate some preservative action). The cooled wort is then inoculated with the yeast.

After fermentation, downstream processing includes the separation of the yeast cells from the beer, and may include a pasteurization stage to inactivate traces of cells passing through the separation stage which can grow and cloud the beer after storage.

Mixed Species Processes

CHEESE MAKING

The use of mixed cultures of *Lactobacilli* species to produce lactic acid (together with milk-clotting enzymes like calf rennet) which initially produce the curds is a well-established practice dating back to the Greek civilization and earlier. Modern cheese processes tend to use a mixture of natural rennet and microbial rennet produced using *Mucor miehei*. Other bacteria, moulds, etc. are also added to the separated curds and incorporated into the cheese block to allow them to act during the ripening stages. These secondary organisms give special characteristics and flavours, and include the use of *Streptococci*, *Penicillia*, *Candida*, etc. to give highly flavoured and/or coloured products.

OTHER FOODS[11]

Pickles, sauerkraut, sourdough bread, fermented meats (wursts), and other food materials are deliberately inoculated with mixed species of micro-organisms to give the distinctive flavours associated with these foods. This practice is a world-wide phenomenon, and products range from fermented fish (Japan), cabbage and other vegetables (Korea, Europe, N. America) to fermented maize (W. Africa). An advantage of the fermentation process is in the production of organic acids (acetic, lactic) which act as preservatives giving the product a longer shelf life.

METHANE PRODUCTION[12, 13]

Mixed cultures can be used anaerobically to produce methane from humus, domestic, farm, and industrial wastes, or from cells deliberately grown as a substrate for the process (biomass). A great deal of research is currently being undertaken in this area (particularly in developing countries) where methane production is seen as a potential source of renewable energy supplies.

EFFLUENT TREATMENT[14, 15]

In order to reduce industrial liquid wastes to a condition suitable for discharge into rivers or estuaries, a variety of compounds (depending on the effluent composition) must be reduced to low concentrations. The compounds in question can be carbohydrates, proteins, organic, and inorganic chemicals. Since these compounds are substrates for a wide range of micro-organisms, it is logical to treat any particular effluent with cultures of mixed population specifically developed for that effluent mixture. Examples of effluent treatment processes are covered in more detail in Chapter 9.

References

1. M. Dixon and E.C. Webb, *Enzymes*, 3rd edn, Academic Press, New York (1979).
2. T. Godfrey and J. Ritchett, *Industrial Enzymology*, Macmillan, London (1983).
3. F.A. Lowenheim and M.K. Norman (eds), *Faith, Keyes & Clark's Industrial Chemicals*, 4th edn, Wiley, New York (1975).
4. H. Ebner and H. Follmann, 'Vinegar', in *Biotechnology* (eds H.J. Rehm and G. Reed), Vol. 5, *Food and Food Production with Microorganisms* (Vol. ed. G. Reed), p. 425, Verlag Chemie, Weinheim (1983).
5. C.H. Holten, A. Muller, and D. Rehbinder, *Lactic Acid*, Verlag Chemie, Weinheim (1971).

6. A. Fiechter, O. Kappell, and F. Meussdoerfer, 'Batch and continuous culture', in *The Yeasts* (eds A.H. Rose and J.S. Harrison), Vol. 2, 2nd edn, Academic Press, London (1986).

7. G.L. Solomons, 'Single cell protein', in *General Review in Biotechnology* (eds C. Stewart and I. Russel), CRC Press, Boca Raton (1983).

8. J. Olsen and K. Allermann, 'Microbial biomass as a protein source', in *Basic Biotechnology* (eds J. Bu'Lock and B. Kristiansen), Academic Press, London (1987).

9. J.S. Hough, *The Biotechnology of Malting and Brewing*, Cambridge University Press (1985).

10. S. Lafon-Lafourcade, 'Wine and brandy', in *Biotechnology*, (eds H.J. Rehm and G. Reed), Vol. 5, *Food and Food Production with Microorganisms* (Vol. ed. G. Reed), p. 81, Verlag Chemie, Weinheim (1983).

11. G. Campbell-Platt, *Fermented Foods of the World—A Dictionary and Guide*, Butterworths, London (1987).

12. W. Palz, P. Chartier and D.O. Hall (eds), *Energy from Biomass*, Applied Science, Barking (1981).

13. D.A. Stafford, D.I. Hawkins, and R. Horton, *Methane Production from Waste Organic Matter*, CRC Press, Boca Raton (1980).

14. M.E. Bushell and J.H. Slater, *Mixed Culture Fermentations*, Academic Press (for Society for General Microbiology), London (1981).

15. IChemE Symposium Series No. 77, *Effluent Treatment in the Process Industries*, Institution of Chemical Engineers, Rugby (1983).

Chapter 2

Sterilization Processes

The aim of an industrial sterilization process is to provide a substrate for the micro-organism to be cultivated which effectively behaves aseptically. If the process is aerobic, it is also necessary to supply air which also behaves aseptically. In practice, this usually means the reduction of naturally occurring organisms in the substrate and air supply to a level where they are unable to compete effectively for the nutrients in the substrate with the organism being deliberately cultivated.

Micro-organisms can be removed from liquids or gases by any of the following methods:[1-3]

- application of heat
- use of chemical agents or irradiation
- mechanical methods—filtration, centrifugation.

The first two methods do not remove the cells, but render them non-viable (i.e. incapable of competing for the nutrients).

Liquid Sterilization

APPLICATION OF HEAT

Thermal methods used for the sterilization of large volumes of liquids are designed to reduce the viability of competing micro-organisms, and the process must be based on a knowledge of the kinetics of the thermal death of micro-organisms.

Thermal Death Kinetics

The destruction of micro-organisms by heat at a fixed temperature (isothermal) follows a first-order reaction,[1-3] the rate of destruction being given by

Table 2.1 Typical values for the constants in Equation (2.1)

	E (*kcal/mol*)	A (*min^{-1}*)
B. stearothermophilus (FS1518)	67.48	4.93×10^{37}
B. subtilis (FS5230)	68.7	9.50×10^{37}
Cl. sporogenes (PA3679)	68.7	1.66×10^{38}
Vegetative cells	20 max	1.20×10^{21}

$$dN/d\theta = -kN$$

where N is the number of organisms at time θ and k is the specific reaction (death) rate constant. Rearranging, we have

$$dN/N = -k\,d\theta$$

and integrating between N_0 (the number of organisms initially) at time 0, and N at time θ gives

$$\ln(N/N_0) = -k\theta$$

In microbiology, the term 'decimal reduction time' (D) is often used, and is the time taken at a particular temperature to reduce the initial population by a factor of 10. Thus

$$N/N_0 = 1/10$$

and

$$\ln(1/10) = -kD$$

So the decimal reduction time (D) is related to the specific death rate constant (k) by

$$D = 2.303/k$$

Spores of micro-organisms are far more difficult to destroy than vegetative cells, and the specific reaction rate constant used for designing a liquid sterilization process must apply to the viable spores and not to viable cells.

The specific death rate constant (k) varies with temperature, and follows a relationship of the Arrhenius type:

$$k = A \exp(-E/RT) \tag{2.1}$$

where A is a constant, E is the 'activation' energy, R is the universal gas constant, and T is the absolute temperature.

Typical values for the constants in this relationship are given in Table 2.1.[1, 4, 5]

For a fixed temperature, if we decide on the ratio of destruction (N_0/N), knowing k, the time which the liquid must be held at that temperature is readily obtained.

Example 2.1

For the inactivation of *B. subtilis* spores, $A = 9.5 \times 10^{37}$ min^{-1}, and $E = 68.7$ kcal/mol. Assuming that a liquid containing these spores is instantaneously sterilized at 115 °C, calculate the time required to give a destruction ratio of 10^6.

Answer

At 115 °C, $k = 9.5 \times 10^{37} \exp\left(-68\,700/(1.9872 \times 388)\right)$ therefore,

$$k = 0.1922, \qquad \ln(N_0/N) = k\theta, \qquad \text{and} \quad \ln(10^6) = 0.1922\,\theta$$

$$\text{thus } \theta = 13.83/0.1922 = 71.9 \text{ min}$$

There are two ways, however, in which a sterilization process can be carried out: batch sterilization and continuous sterilization.

Batch Sterilization

The medium is prepared and pumped to an agitated vessel fitted with either a heating jacket or heating coils. The contents of the vessel are then heated to a particular sterilization temperature (usually 121 °C), held for a certain length of time at that temperature, and then cooled down to the fermentation temperature. This type of sterilization process is frequently carried out in the fermentor itself.

Continuous Sterilization

The medium is pumped at a constant flowrate through two heat exchangers. One heats the liquid and the other cools it down. In between the two heat exchangers, the liquid is held either in a vessel or a large volumetric capacity pipeline to give the required time at the sterilizing temperature.[6]

In both cases, there will be some parts of the process where the temperature is varying and hence k is varying. Typical temperature profiles for both types of process are shown in Fig. 2.1.

From the basic rate equation for the thermal death of the organism in periods A and C in Fig. 2.1,

$$dN/N = -k\,d\theta$$

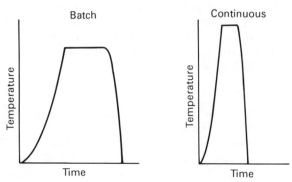

Fig. 2.1 Temperature profiles during liquid sterilization.

so

$$\ln(\mathcal{N}_0/\mathcal{N}) = \int k \, d\theta$$

where k is now a function of time because the temperature is varying with time. Thus:

$$\ln(\mathcal{N}_0/\mathcal{N}) = \int A \exp(-E/RT) \, d\theta$$

In order to design a sterilization process to give a desired result, Deindoerfer and Humphrey[5] defined a term ∇ which is a measure of the destruction ratio $(\mathcal{N}_0/\mathcal{N})$:

$$\nabla = \ln(\mathcal{N}_0/\mathcal{N})$$

This term ∇ is used as a design criterion in order to ensure that different sizes of batch (or different flowrates in continuous sterilizations) all receive the same thermal treatment.

Deindoerfer and Humphrey assumed that the contribution of the sterilizing effect in each stage of the process is additive, so that

$$\nabla_{total} = \nabla_{heating} + \nabla_{holding} + \nabla_{cooling}$$

If we know the temperature profile of a sterilization process, we can calculate the value of k for the different times during the process and plot these values of k against time. Graphical integration will then give the value of ∇_{total} for that particular process.

The contribution to the sterilization at temperatures below $100\,^{\circ}\mathrm{C}$ is minimal, and in practice is ignored.

Example 2.2
10 000 L of medium is sterilized in a fermentor at a temperature of $120\,^{\circ}\mathrm{C}$. The time/temperature profile of the sterilizing process from $100\,^{\circ}\mathrm{C}$ is given in Table 2.2.
Assuming that the contaminating species is *B. stearothermophilus*

(a) calculate the design criterion ∇ for the process
(b) what will be the destruction ratio $\mathcal{N}_0/\mathcal{N}$?

Answer
(a) For each temperature a value of k can be calculated, based on the values for A and E given in Table 2.1. Thus, values of k for different times can be compiled as shown in Table 2.2, and plotted as in Fig. 2.2 (see p. 16).

Graphical integration (counting squares, Simpson's rule, etc.) of the plot in Fig. 2.2 gives a value of 34.0. Thus

$$\nabla_{total} = 34.0$$

(b) Since $\ln(\mathcal{N}_0/\mathcal{N}) = \nabla_{total}$, for this process $\mathcal{N}_0/\mathcal{N} = 5.84 \times 10^{14}$.

The importance of calculating ∇ and of specifying value for ∇ is that based on practical results, if we ensure that the medium always receives the heat treatment corresponding to the design value of ∇, then we will always ensure an acceptable

Table 2.2   Time/temperature profile—Example 2.2

Temperature	Time	k
100	54	0.0143
105	61	0.0477
110	69	0.154
115	79	0.483
120	91	1.47
120	101	1.47
115	104	0.483
110	107	0.154
100	114	0.0143

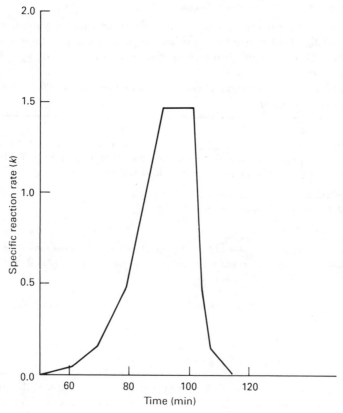

Fig. 2.2   Plot of *k vs* time for Example 2.2.

aseptic medium before inoculation, *irrespective of the size of batch or flowrate of substrate*.

The determination of the value of V to be used can be carried out in a number of ways.

Experimental Determination

If laboratory experiments are carried out using different sterilizing holding times until an acceptable aseptic medium is obtained (i.e. a substrate subsequently proved suitable for the desired fermentation), the acceptable process can be analysed to evaluate the design criterion V. This value of V can then be used for all sizes of batch (or flowrates in continuous sterilizations).

Theoretical Determination

Use can be made of recommended values for V published in the literature to design the sterilization process. If data for the contaminating organism is not known, that for *B. stearothermophilus* may be used.

In practice, values of V between 30 and 80 are used. Deindoerfer and Humphrey suggested that a value of 40 is suitable for most industrial fermentations.

Since $V = \ln(N_0/N)$,

$V = 30$ corresponds to a ratio of $(N_0/N) = 1.0 \times 10^{13}$
$V = 40$ corresponds to a ratio of $(N_0/N) = 1.0 \times 10^{17}$
$V = 80$ corresponds to a ratio of $(N_0/N) = 1.0 \times 10^{34}$

In the food industry, the criterion for sterility in canning operations is the probability of the survival of 1 spore in 10^{12} of *Cl. botulinum* (corresponding to a value for V of 27.6).

Because the graphical integration method is tedious and time consuming, there have been many attempts to simplify the procedure, and that suggested by Richards[7] has found most favour.

Richards assumed a 'standard' heating and cooling curve above 100 °C to be a straight line with a slope of 1 °C/min. He then constructed a table of values of k and cumulative V values for 1 °C incremental steps. This allows a rapid calculation of the value of V_{total} for any process, where only the sterilizing temperature (holding temperature) and the time required to reach this temperature from 100 °C (and to cool from this temperature back to 100 °C) need be known.

Example 2.3

In Example 2.2 above, the times taken for the heating, holding and cooling periods are as follows:

heating from 100 °C to 120 °C   37 min
cooling from 120 °C to 100 °C   13min
holding at 120 °C   10 min.

Using the Richards method, calculate the value of V_{total} for this process.

Table 2.3 Table constructed using data for *B. stearothermophilus.*
$A = 4.93 \times 10^{37}$; $E = 67.48$ kcal/mol

Temperature (°C)	*k*	*Cumulative* ∇
100	0.0143	—
101	0.0182	0.0325
102	0.0232	0.0558
103	0.0296	0.0854
104	0.0376	0.1229
105	0.0477	0.171
106	0.0604	0.231
107	0.0765	0.308
108	0.0967	0.404
109	0.122	0.526
110	0.154	0.681
111	0.194	0.875
112	0.244	1.12
113	0.307	1.43
114	0.385	1.81
115	0.483	2.29
116	0.605	2.90
117	0.757	3.66
118	0.945	4.60
119	1.18	5.78
120	1.47	7.25
121	1.83	9.08
122	2.28	11.36
123	2.83	14.19
124	3.51	17.70
125	4.35	22.05
126	5.39	27.45
127	6.67	34.11
128	8.24	42.36
129	10.18	52.54
130	12.55	65.08

Answer

(a) *Heating* From Table 2.3, the cumulative values of ∇ (100 °C to 120 °C) is 7.25. Since the actual process time is 37 min and not 20 min (assumed in preparing Table 2.3), more sterilization will have taken place, and:

$$\nabla_{heating} = 7.25 \times 37/20 = 13.41$$

(b) *Cooling* From Table 2.3, ∇ for a 20 min cooling time is 7.25. Since the process time for cooling is less (13 min), less sterilization will have taken place:

$$\nabla_{cooling} = 7.25 \times 37/20 = 4.71$$

(c) *Holding*   This part of the process is isothermal for 10 min, and $\nabla_{\text{holding}} = k\theta$.
   From Table 2.3, at 120 °C $k = 1.47$

$$\nabla_{\text{holding}} = 1.47 \times 10 = 14.7$$

For the total process,

$$\nabla_{\text{total}} = 13.41 + 4.71 + 14.7 = 32.82$$

This compares with a value for ∇_{total} of 34.0 calculated in Example 2.2 using graphical integration—an error of 3.5 per cent.

In general, the use of the Richards method will give an answer to within ± 5 per cent of the rigorous method.

One of the advantages of the Richards method is that adjustments can be quickly made to the process in the case of malfunctions (for example, of the control system).

Example 2.4
During the batch sterilization of a liquid, normally carried out at 121 °C for 10 min, the control system showed a malfunction when the temperature reached 116 °C. Due to the malfunction, the temperature remained at 116 °C for 15 min before the fault was rectified. What new holding time at 121 °C will be required to ensure that the design criterion ∇ is maintained for this batch?

Answer
Normally,

$$\nabla_{\text{total}} = \nabla_{\text{heating}} + \nabla_{\text{holding}} + \nabla_{\text{cooling}}$$

For the process with the control malfunction:

$$\nabla_{\text{total}} = \nabla_A + \nabla_B + \nabla_C + \nabla_N + \nabla_{\text{cooling}}$$

where:

   ∇_A = heating from 100 °C to 116 °C (normal pattern)
   ∇_B = holding at 116 °C for 15 min (during malfunction)
   ∇_C = heating from 116 °C to 121 °C (normal pattern)
   ∇_N = new holding period at 121 °C.

Since for the normal operation $\nabla_{\text{heating}} = \nabla_A + \nabla_C$ and ∇_{cooling} is the same for both cases,

$$\nabla_{\text{holding}} = \nabla_B + \nabla_N$$

At 116 °C, $k = 0.605$ and $\nabla_B = 0.605 \times 15 = 9.08$.
   At 121 °C, $k = 1.83$ and (normally) $\nabla_{\text{holding}} = 1.83 \times 10 = 18.3$, thus

$$\nabla_N = (18.3 - 9.08) = 9.22$$

and the new holding time is

$$9.22/1.83 = 5 \text{ min}$$

CHEMICALS AND IRRADIATION[3]

For the sterilization of liquids, it is possible to use disinfecting chemicals, e.g. phenols, ethylene oxide, hypochlorites, but the main problem is that these must be used in excess, and the medium not only becomes sterile in terms of reducing the population of competing organisms, but will also kill off some of the organisms which are subsequently added by inoculation. Thus the use of chemical sterilants is not generally suitable for the sterilization of bulk liquid quantities of medium.

Irradiation (using ultraviolet, X-rays, gamma radiation) can be used to sterilize liquids, but in the volumes required for industrial *batch* processes, the irradiation equipment would be massive in size, and could well represent a major safety hazard.

Continuous sterilization of substrate is however now possible, since the liquid flows continuously through a small ionizing radiation chamber, and the control and safety of the equipment is satisfactory. Since the process is in its infancy, details of operating economics are difficult to obtain. It is claimed, however, that the economics are as favourable as thermal sterilization for moderate sizes of fermentor.

MECHANICAL METHODS

Microbiological filters capable of removing micro-organisms from liquids are available for both laboratory and full-scale use. However, for the volumes of liquids required for use as a fermentor substrate, this method of sterilization is uneconomic compared to thermal methods.

Biological filters are used, however, for the sterilization of liquid *products* after recovery, where the volumes of liquid involved are very much smaller than at the fermentation stage.

Centrifugation as a method of removing competing organisms on a large scale is an uneconomic process.

Sterilization of Gases

As in the case of sterilization of liquid media, the main aim in the sterilization of gases is to reduce the viable organism population to an acceptable, non-competing level. Also, as in the case of liquid sterilization, removal of organisms can be carried out in a number of ways:[1-3]

- mechanical collection (e.g. filtration)
- use of chemicals or irradiation
- application of heat.

The major method used for the removal of micro-organisms from the large volumes of air required for aerobic fermentations is in the use of fibre filters, although other methods do have a limited use.

FILTRATION OF GASES[8, 9]

Since micro-organisms are of small physical size ($0.5–2\,\mu$m wide $\times\ 0.5–2\,\mu$m long), the separation mechanism used for the removal of organisms from flowing gases must be different from the mechanism used for the separation of solid particles from a liquid suspension.

In the filtration of a solid from a liquid suspension, the filter cloth (or paper) is chosen such that the pores in the filter cloth are smaller in size than the solid, and basically the separation takes place on the surface of the cloth and on the bed of particles formed (see p. 73).

If this method were to be used for the removal of micro-organisms, the pore size of the filter would have to be smaller than the organism ($0.5–2.0\,\mu$m). For large volumetric flowrates of gas, the high pressure drop across the filter would impose an uneconomic use of energy due to the pumping costs.

The mechanism used for the removal of micro-organisms from gases is to trap the organism *inside* the filter medium (not on the surface). The type of filter used is normally a fibrous mat, relatively thick, consisting of fibres of material fabricated in random fashion so that the gas must follow a tortuous path in order to pass from one side of the filter to the other (see Fig. 2.3). Collection of the organisms takes place by collision of the organism with one of the fibres of the mat.

Fibrous filters are dependent on the velocity of the gas for their collection efficiency, the variation of retention efficiency with gas velocity being similar to that shown in Fig. 2.4. The minimum value of N_0/N for bacterial spores occurs at a gas velocity of approximately 0.3 m/s. Above this critical value there is a dramatic increase in the collection efficiency (the higher the velocity the more chance of collision of the organisms with the fibres).

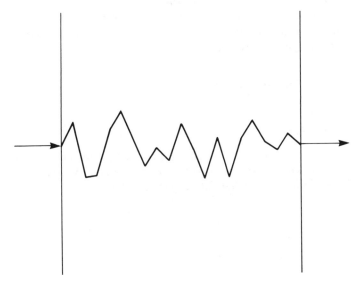

Fig. 2.3 Path of gas (and organisms) through a fibre filter.

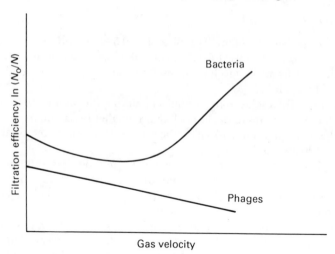

Fig. 2.4   Filtration characteristics of a fibre filter.

For phages, where the size (and hence mass) is much smaller even than for spores, the gas flow must be kept low to improve the chances of collection, otherwise the lighter phages are carried past the fibres with the air stream.

The removal efficiency N_0/N is related to the mat thickness by the relationship

$$\ln(N_0/N) = KL$$

where L is the filter thickness and K is a filtration constant.

The filtration constant (K) is a function of gas velocity, fibre size, and the density of the organism to be removed, and it is normal practice to express K in terms of the filter thickness required to remove 90 per cent of the organisms in the gas (L_{90}):

$$\ln(N_0/N) = \ln(100/10) = KL_{90}$$

and

$$K = 2.303/L_{90}$$

Knowing L_{90} for a particular material (fibre) at a particular gas velocity will enable a filter thickness to be specified for *any* value of N_0/N. Typical collection characteristics for fibre mat filters are given in Table 2.4.

Example 2.5
For the removal of *B. subtilis* spores from an air stream flowing at 1.54 m/s through a filter composed of 16 μm glass fibre, what thickness of filter will be required for removal ratios of:

(a) 1000:1, (b) 10 000:1, (c) $10^{12}:1$ and (d) $2.3 \times 10^{17}:1$?

Answer

For this type of filter at a gas velocity of 1.54 m/s, $L_{90} = 1.52$ cm (Table 2.4).
 Since the filter constant $K = 2.303/L_{90}$
$$K = 1.515$$
and $\ln(N_0/N) = 1.515\ L$

(a) $\ln(1000)$ $= 1.515\ L$: $L = 6.908/1.515 = 4.56$ cm
(b) $\ln(10\,000)$ $= 1.515\ L$: $L = 6.08$ cm
(c) $\ln(10^{12})$ $= 1.515\ L$: $L = 18.24$ cm
(d) $\ln(2.3 \times 10^{17}) = 1.515\ L$: $L = 26.4$ cm

Table 2.4   Characteristics of fibre mat filters

Filter material	Fibre size	Test organism	Gas velocity	L_{90} (cm)
Glass fibre	16 μm	B. subtilis	0.031	4.10
		spores	0.154	9.14
			0.310	11.70
			1.54	1.52
			3.08	0.38
Glass fibre	8.5 μm	E. coli	0.031	0.43
		phage	0.154	0.61
			0.310	0.71
			1.54	0.86
			3.08	1.12

A wide range of commercial filters are available for the removal of organisms from gas streams. Fibre mats manufactured from glass fibre and rock wool (mineral fibre) are available, and it is possible to obtain mats of glass wool (much finer than spun glass fibre).

Materials which 'wet' easily (mainly natural fibres like cotton and wool) are to be avoided, because if they are wetted, the size of pores through the material tend to be enlarged on drying, and the reproducibility of collection efficiency suffers.

Newer materials such as plastic filters (polyethylene, PTFE, PVC) are also in use, although problems can arise in sterilization of the filter due to the relatively low softening temperature. Sintered metals (stainless steel) can also be used.

The major engineering problem is that the placing of a filter in the air flow channel causes a pressure drop. The denser the fibre mat, the greater the resistance to the flow of air—thus the cost of pumping the air through the filter rises, and an economic balance must be achieved between the cost of the filter and the cost of pumping. For example, a 30 cm mat of glass fibre or rock wool may be more economic overall than 7 cm of sintered stainless steel (5–6 μm aperture).

Filtration of organisms does not of course destroy the organisms, and after a long period of operation some organisms will penetrate the filter because successive layers of fibre can collect no more. Filters must therefore be capable of being sterilized, and this generally means the use of thermal methods.

Filters can be sterilized in two ways:

- treatment with *'live' steam* (121 °C for up to 30 min) will ensure that the majority of the trapped spores will be inactivated, but the filter material must be capable of withstanding the temperature, and a number of plastics cannot tolerate 121 °C. Cotton and wool (natural) are also wetted and are not suitable for sterilization by this method.
- *'dry' heat* can also be used (using electrical heating elements), and this process (in the presence of oxygen) is dealt with more fully below.

CHEMICAL AGENTS

It is possible to destroy the activity of micro-organisms by the use of disinfectant solutions, and in the laboratory sufficient sterility can be maintained by bubbling air through such a solution, using for example a Drechsel bottle. Little work has been carried out on using this method of air (gas) sterilization on an industrial scale, but there is no reason to believe that this would not be successful since efficient gas/liquid contacting devices are in wide use throughout the process industries. Obviously, to compete with other available methods, the economics of the process would have to be favourable.

APPLICATION OF HEAT

We have already seen that spores of micro-organisms can be killed using heat, and that the rate of destruction follows a first-order reaction:

$$\ln(N_0/N) = k\theta$$

The specific reaction rate constant (k) is also a function of temperature:

$$k = A \exp\left(-E/RT\right)$$

The sterilization of culture medium is a 'wet' heat process, the mechanism of destruction being due to protein destruction or denaturation.[2, 3]

For the sterilization of gases it is possible to use a 'dry' heat process. For 'dry' heat destruction it is essential that oxygen is present because the destruction mechanism is one of oxidation. It is also essential that the air is not stagnant, since a continuous replenishment of oxygen to the organism is necessary.[10] For this type of 'dry' process,

$$E_d = 12\,000 – 24\,000 \text{ cal/mol}$$

For a 'wet' process,

$$E_w = 40\,000 – 80\,000 \text{ cal/mol}.$$

The value of A in the Arrhenius expression is the same for both 'wet' and 'dry' processes (typically 10^{37} to 10^{39}).

Example 2.6

For the destruction of *B. stearothermophilus* spores using both 'wet' and 'dry' processes, the following values apply in the Arrhenius equation:

$$A = 4.93 \times 10^{37} \text{ (both 'wet' and 'dry')}$$
$$E_d = 24 \text{ kcal/mol ('dry' oxidation process)}$$
$$E_w = 67.5 \text{ kcal/mol ('wet' process)}$$

Calculate the value of the specific reaction rate for both processes at 105 °C.

Answer

(a) 'Wet' process:

$$k = A \exp(-E_w/RT) = 4.93 \times 10^{37} \exp(-67\,500/1.987(273+105))$$
$$= 0.0477 \text{ min}^{-1}$$

(b) 'Dry' process:

$$k = 4.93 \times 10^{37} \exp(-24\,000/1.987(273+105))$$
$$= 6.6 \times 10^{23} \text{ min}^{-1}$$

It can be seen from Example 2.6 that a dry process (in the presence of oxygen) would require either a short contact time (seconds only) or the use of a much lower temperature.

The use of electrical heating for sterilization of large volumes of air has not received much attention, and dry heat is mainly limited to the sterilization of fibre filters where the normal practice is to sterilize the filter and its associated pipework using flowing air heated to between 80° and 100 °C.

The most common method used for sterilizing air using the dry oxidation process is to use the heat generated by the adiabatic compression of the gas.

If air is compressed from 1 atmosphere to 6.8 atmospheres, the air reaches a temperature of approximately 200 °C at the compressor outlet, and the air can be considered completely sterile at this point. The only disadvantage is that the cost of compression becomes excessive at high gas flowrates due to the high cost of the machine.

For example, a 100 m³ batch fermentor processing *B. subtilis* for α-amylase production may require 100 m³/min air flow. If this air flow is also to be provided at 6–7 atmospheres pressure, the cost of the machine would be prohibitive. This method is therefore only used for relatively small processes (under 10 000 L).

The use of steam for gas sterilization is not favoured due to the poor heat transfer characteristics between the two 'gases'.

Pipeline Sterilization

Irrespective of the method used for the sterilization of the air (or liquid medium), the lines through which the gas passes must also be sterile, or the whole objective of the sterilization process will fail. Pipelines are usually sterilized by passing steam through them, or, in the case of dry filters, by passing heated air through the line.

A typical pipeline sterilization layout is shown in Fig. 2.5.

A summary of notation used in this chapter is shown in Table 2.5.

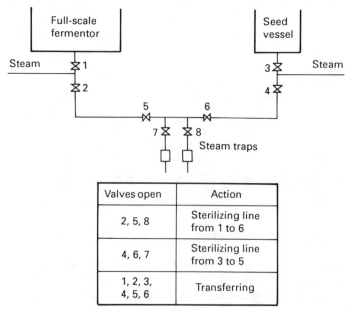

(a) Inoculation line

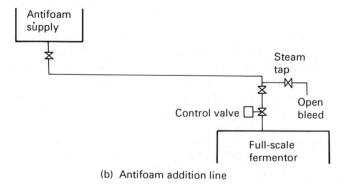

(b) Antifoam addition line

Fig. 2.5   Layout for pipeline sterilization.

References

1. S. Aiba, A.E. Humphreys, and N.F. Millis, *Biochemical Engineering*, 2nd edn, University of Tokyo Press (1979).
2. A.D. Russell, *The Destruction of Bacterial Spores*, Academic Press, London (1982).
3. L.B. Quesnel, 'Sterilisation and sterility', in *Basic Biotechnology* (eds J. Bu'Lock and B. Kristiansen), Academic Press, London (1987).

Table 2.5 Summary of notation used in Chapter 2

A	constant in Arrhenius expression	min^{-1}; s^{-1}
D	decimal reduction time	s; min
E	'activation' energy (Arrhenius expression)	kcal/mol
E_d	'activation' energy for dry oxidation	kcal/mol
E_w	'activation' energy for wet process	kcal/mol
k	specific destruction rate	min^{-1}; s^{-1}
K	filter constant for fibre mat	m^{-1}
L	filter thickness	m
L_{90}	thickness to remove 90 per cent of organisms	m
$\mathcal{N}$	number of viable spores	no/mL; no/m^3
$\mathcal{N}_0$	initial spore population	no/mL; no/m^3
R	Universal gas constant	kcal/mol K
T	absolute temperature	kelvin
∇	sterilization criterion $[=\ln(\mathcal{N}_0/\mathcal{N})]$	—
θ	time	s; min

4. F.H. Deindoerfer and A. Humphrey, 'Calculation of heat sterilisation times for fermentation media', *Appl. Microbiol.* **5**, 221 (1957).
5. F.H. Deindoerfer and A. Humphrey, 'Analytical methods for calculating heat sterilisation times', *Appl. Microbiol.* **7**, 256 (1959).
6. M.H.J. Ashley, 'Continuous sterilization of media', *The Chemical Engineer, London*, No. 377, 54 (1982).
7. J.W. Richards, 'Rapid calculations for heat sterilisations', *Br. Chem. Eng.* **10**, 166 (1965).
8. S.K. Friedlander, 'Aerosol filtration by fibrous filters', in *Biochemical and Biological Engineering Science* (ed. N. Blakebrough), Vol. 1, p. 49, Academic Press, London (1967).
9. J.W. Richards, 'Air sterilisation with fibrous filters', *Proc. Biochem.* **2**(9), 21 (1967).
10. G.L. Solomons, *Materials and Methods in Fermentation*, p. 79, Academic Press, London (1969).

Chapter 3

Fermentor (Reactor) Design Theory

In designing a fermentor to carry out a particular operation, a number of factors are important in establishing a satisfactory end result, i.e. a piece of equipment which will give the required performance.

The first stage in the design of a fermentor vessel is to estimate the physical size of the equipment. In order to do this, the decision must be taken whether to operate on a batch basis with a single fermentor, a batch basis using multiple fermentors, or to operate continuously.

In order to obtain the size of the vessel, the rate of production is important, and essential information on the growth rate of the micro-organism plays a major part in the prediction of the time cycle of a batch fermentation.

The estimation of the cycle time can be made using kinetic models of the growth rate when matched to the desired production rate. This time cycle can then be used to determine the fermentor size.

In order to calculate the size of the vessel, the availability of the vessel for processing must also be known; for example, a single fermentor used for the same production rate working an eight-hour day, five days a week will have to be much larger than a single fermentor giving the same production rate but working a 168 hour, seven-day week.

The physical size of the fermentor vessel cannot be considered in isolation from other factors such as heat transfer and agitation characteristics, and the final design must be dictated by economic considerations (taking into account both capital and operating costs). These factors are dealt with in more detail in Chapter 4.

This chapter deals with the use of kinetic growth data of micro-organisms and the models which can be used to predict the time of the different phases which occur during growth in order to determine the overall time cycle for the fermentation. This information can then be used to give an initial guide to the size of the fermentor before proceeding to consider the other factors involved.

The same models used for the design of a batch process can also be used to predict the size of fermentor required for continuous fermentation processes, and examples of the use of these models for continuous fermentation are also included in this chapter.

Cell Growth and Production Kinetics

The general aim in a biological reaction is to support the growth of a specific organism and to encourage a high product yield. This does not necessarily mean that we must provide all essential nutrients in a large excess, because in certain circumstances excessive concentrations of nutrients can inhibit or poison cell growth. It is therefore common practice to limit the concentration of at least one of the essential nutrients to give a controlled overall growth.

If the concentration of one essential nutrient is varied, keeping all other nutrients constant and in excess, the growth rate is found to vary exponentially as shown in Fig. 3.1. For unicellular growth, the growth rate can be expressed in terms of cell concentration X, and the specific growth rate μ is defined by:

$$\mu = \frac{1}{X} \frac{dX}{d\theta}$$

where θ is time.

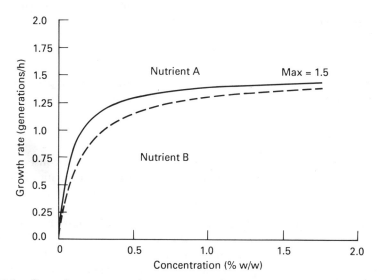

Fig. 3.1 Growth rate *vs* nutrient concentration.

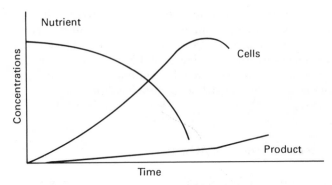

Fig. 3.2   Growth and production rates during batch processing.

If the value of μ is constant, this represents exponential growth of the organism (proportional to the cells present).

The growth rate of cells (and usually the rate of synthesis of a desired product) in a batch fermentation will be similar to that shown in Fig. 3.2.

As the cells grow, the population shows an exponential increase until the concentration of limiting nutrient reaches a low level, at which stage cells start to die off. Assuming that the desired product is formed at the same rate as cells, this is the time to stop the fermentor batch and prepare to start a new one.

In the case where product formation lags behind cell production, the batch is allowed to continue (even though cells are dying) until the optimum concentration of product is achieved.

In the laboratory the 'mass doubling time' (θ_d) is frequently measured, and the specific growth rate μ for unicellular growth is related to the mass doubling time by

$$\theta_d = \ln(2)/\mu$$

For the case of a limiting substrate nutrient, the relationship proposed by Monod[1] is widely used:

$$\mu = \frac{\mu_m C_i}{K_i + C_i}$$

where μ_m is the maximum specific growth rate, K_i is a 'saturation' constant, and C_i is the concentration of limiting nutrient i. μ_m is the maximum specific growth rate measured when $C_i \gg K_i$, and K_i is the value of the concentration of nutrient i when $\mu = \mu_m/2$.

The relationship proposed by Monod was derived on an empirical basis and is a simplified model of the complicated growth pattern. Determination of μ_m and K_i in the laboratory will give a measure of the population growth, which may be used to predict the cell population characteristics in a large-scale fermentor, particularly if nutrient uptake is also measured.

Where the situation is more complex, e.g. two limiting nutrients, toxin formation, inhibition by product concentration, various modifications can be made to the Monod expression[2] to determine the specific growth rate μ at any concentration, and the modified equation can be used simultaneously to predict the maximum cell population (see pp. 34–5 below).

- *Two limiting nutrients:*

$$\mu = \mu_m \left(\frac{C_1}{C_1 + K_1} \right) \left(\frac{C_2}{C_2 + K_2} \right)$$

- *Product inhibition:*[3]

$$\mu = \mu_m \left(\frac{C_i}{C_i + K_i} \right) \left(\frac{C_P}{C_P + K_P} \right)$$

where K_P is the 'saturation' constant for the product measured as K_i, and C_P is the concentration of the product.

An example of product inhibition is the glucose/yeast/alcohol fermentation.

Batch Fermentation

A typical batch fermentation process can be schematically illustrated as in Fig. 3.3, where the growth of an organism normally shows four distinct phases:

- lag phase
- exponential growth phase
- stagnant phase (maximum population)
- death phase.

THE LAG PHASE

Batch culture always involves the inoculation of a sterile medium with a prepared culture of the micro-organism to be grown. After inoculation, apart from air addition to aerobic processes and exhaustion of gases produced during the fermentation, nothing else is added or extracted from the batch.

The length of the lag phase depends on the changes in nutrient composition experienced by the organism on inoculation, and on the age of the inoculum.

The growth of cells depends on many factors, and the rapid change to a new environment (the sterile fermentor) can affect several important variables:

- inoculation into a medium with higher concentrations of nutrients can cause a delay in cell growth until the organism adapts to its new environment
- necessary molecules synthesized by the cell to promote growth (vitamins, activators) may be lost by diffusion out of the cell due to a dilution effect, and may take time to be replenished by the cells
- the size of the inoculum and number of viable cells play an important part in the lag phase

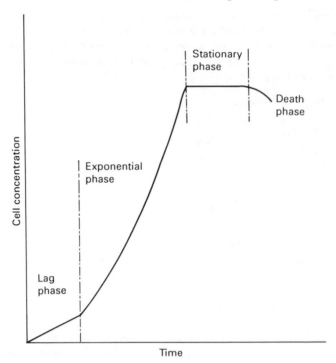

Fig. 3.3   Batch fermentation phases.

- the age of the inoculum is also important because new cells will not have produced the same quantity of metabolic intermediates as cells already in their exponential growth phase.

If we assume that the lag phase ends once a certain important component *in the cell* reaches a critical value (c), then:

$$c = aV + bN_0\theta_l + d\theta_l$$

where V is the volume of inoculum, N_0 is the number of cells/new volume, a is the critical substance concentration/old volume $\times$ (old : new volume), b is the increase in critical substance/time per cell (old cells), d is the internal cell production (new cells), and θ_l is the time of the lag phase.

So

$$\theta_i = \left(\frac{c}{b} - \frac{aV}{b}\right)\Big/\left(N_0 + \frac{d}{b}\right)$$

and it can be seen that θ_l is dependent on V, so the larger V is, the shorter will be θ_l.

It can also be seen that for large volumes of inoculum, θ_l is proportional to $1/N_0$. Process design of the fermentation part of the overall process should aim to minimize the length of the lag phase in order to obtain maximum utilization of the fermentor, and the following points are of importance:

- the inoculum should be as active as possible (preferably in the exponential growth phase)
- the medium used to grow the inoculum should correspond as closely as possible to the medium to be used in the full-scale fermentor
- a reasonably large volume of inoculum should be used to minimize loss by diffusion of metabolic intermediates (at least 5 per cent of the fermentor volume).

Example 3.1
Assuming a minimum 5 per cent inoculum volume, what size of laboratory fermentor would be required to seed a 10 000 L full-scale fermentor?

Answer
 5 per cent of 10 000 L is 500 L
 5 per cent of 500 L is 25 L
 5 per cent of 25 L is 1.25 L

Thus, 1.25 L of laboratory culture would be required to seed a 25 L fermentor. This would be used to seed a 500 L fermentor which in turn would be used to seed the 10 000 L full-scale fermentor.

EXPONENTIAL GROWTH PHASE

The exponential growth rate is given by:

$$\mu = \frac{1}{X}\frac{dX}{d\theta}$$

or on integration

$$\ln\left(\frac{X}{X_0}\right) = \mu\theta$$

where X_0 is the initial population at time 0 and X is the population at time θ.
 For a limiting nutrient the Monod expression also gives μ, and:

$$\mu = \frac{\mu_m C_i}{C_i + K_i}$$

Thus, combining these two expressions, the cell population X at time θ is given by

$$\ln\left(\frac{X}{X_0}\right) = \frac{\mu_m C_i \theta}{C_i + K_i}$$

and this expression can be used to predict the exponential growth of the organism, if the Monod expression applies.

The Monod equation tends to break down if growth is very rapid, and the following models have had some success in difficult cases:[2]

$$\mu = \frac{\mu_m C_i}{K_i + C_0}$$

where C_0 is the initial concentration of nutrient, or

$$\mu = \frac{\mu_m C_i}{C_i + BN}$$

where B is a constant and N is the cell population.

STATIONARY PHASE AND MAXIMUM POPULATION

Assuming that the rate of consumption of nutrient $(dA/d\theta)$ is proportional to the number of viable cells until we reach the stationary phase, then

$$\frac{dA}{d\theta} = -K_A X \tag{3.1}$$

where X is the number of cells, and K_A is a proportionality constant for nutrient A.
   If the cells follow exponential growth,

$$\mu = \frac{1}{X} \frac{dX}{d\theta}$$

or

$$X = \frac{1}{\mu} \frac{dX}{d\theta} \tag{3.2}$$

Substituting (3.2) in (3.1) gives

$$\frac{dA}{d\theta} = -K_A \frac{1}{\mu} \frac{dX}{d\theta}$$

and

$$dA = -\frac{K_A}{\mu} dX$$

Thus, if the initial concentration of $A = A_0$ at $X = X_0$ (initial concentration of cells), when $A = 0$, $X = X_s$ (the static population), and will be given by

$$\int dA = -\frac{K_A}{\mu} \int dX$$

so

$$0 - A_0 = -\frac{K_A}{\mu} (X_s - X_0)$$

and

$$X_s = X_0 + \frac{\mu}{K_A} A_0$$

This relationship gives the maximum population (X_s) for an initial concentration of nutrient (A_0) and initial cell population (X_0) at the start of the exponential growth phase.

If equation (3.2) is integrated between $X = X_0$ and $X = X_s$, the time taken to reach the stationary phase can then be calculated.

DEATH PHASE

In normal industrial practice, the fermentation is stopped either before this point is reached or just as the cell population starts to decrease as a result of complete utilization of nutrient.

The only exception would be when product formation exhibits a large 'lag' behind cell production, and then, if the product concentration is still rising, the fermentation may proceed past the start of the death phase.

MOULD FERMENTATIONS

All of the above analysis holds true for bacterial fermentations and growth. In the case of mould growth in deep fermentation, the growth rate is much slower than with bacteria, and does not necessarily exhibit exponential growth.

If we assume a spherical pellet growing in submerged culture, then the rate of increase in size of pellet:

$$\frac{dR}{d\theta} = K \qquad \text{(a constant)}$$

The radius will increase at a constant rate in surface culture; for filaments the length will increase at a constant rate.

The mould mass is given by

$$M = (\text{sphere volume} \times \text{density})$$
$$= 4\pi R^3 \rho / 3 \tag{3.3}$$

Also,

$$\begin{bmatrix} \text{rate of} \\ \text{increase} \\ \text{of mass} \end{bmatrix} = \begin{bmatrix} \text{surface} \\ \text{area} \end{bmatrix} \times \begin{bmatrix} \text{increase in} \\ \text{thickness} \\ \text{rate} \end{bmatrix} \times \begin{bmatrix} \text{density} \end{bmatrix}$$

$$\frac{dM}{d\theta} = 4\pi R^2 \frac{dR}{d\theta} \rho \tag{3.4}$$

From (3.3),

$$R = \left(\frac{3M}{4\pi\rho}\right)^{1/3}$$

and substituting in (3.4) for R gives

$$\frac{dM}{d\theta} = 4\pi\rho K\left(\frac{3M}{4\pi\rho}\right)^{2/3} = gM^{2/3}, \qquad (g = K(36\pi\rho)^{1/3} = \text{constant})$$

Integrating between M_0 at time 0 and M at time θ,

$$\int \frac{dM}{M^{2/3}} = g \int d\theta$$

gives

$$3M^{1/3} - 3M_0^{1/3} = g\theta$$

and

$$M = (M_0^{1/3} + z\theta)^3, \qquad (z = g/3)$$

Since $M_0 \ll M$, the mould mass is dependent on $(\text{time})^3$.

The situation with moulds is very complex and various models have been developed in an attempt to model the growth—usually of a modified Monod type.[4]

Continuous Fermentation

The design and analysis of continuous fermentation operation is based on the concept of the continuously stirred tank reactor or fermentor (CSTF). In biotechnology, this type of fermentor is referred to as a *chemostat*.

The assumptions made for the analysis of the CSTF are:[2]

- perfect mixing occurs so that the exit stream has the same composition as the rest of the vessel contents
- mixing is such that the concentration of all components within the vessel is the same in all parts of the vessel
- if the process is aerobic, the concentration of dissolved oxygen is the same in all parts of the vessel
- the heat transfer characteristics of the system are constant (heat of fermentation is removed continuously).

When a steady state has been reached (concentrations, cell population, and temperature at any point in the vessel do not change with time) we can apply a mass balance to any component such that:[5]

$$\begin{bmatrix} \text{rate} \\ \text{of addition} \\ \text{to system} \end{bmatrix} - \begin{bmatrix} \text{rate} \\ \text{of removal} \\ \text{from system} \end{bmatrix} + \begin{bmatrix} \text{rate} \\ \text{of production} \\ \text{within system} \end{bmatrix} = 0$$

CELL GROWTH

If X is the cell concentration in vessel (and in the exit stream), X_0 is the cell concentration in the feed stream, F is the volumetric flowrate of feed stream (and exit stream), V is the vessel volume, r is the rate of cell formation (cells/unit time/unit volume) $(=dX/d\theta$, then a steady-state cell balance (neglecting any deaths) is given by:

$$F(X_0-X)+rV=0$$

or

$$\left(\frac{F}{V}\right)X_0=\left(\frac{F}{V}\right)X-r$$

but the specific growth rate μ is given by

$$\mu=\frac{r}{X}$$

Thus

$$DX_0=X(D-\mu)$$

$D(=F/V)$ is known as the *dilution rate*. The dilution rate (D) is the number of vessel volumes passing through the vessel/unit time (it is the inverse of the mean residence time or holding time). The concept of the dilution rate is used almost universally throughout the biotechnology industry.

For a single continuous fermentor, the feed stream is normally sterile nutrient only, and therefore X_0 is zero, so

$$X(D-\mu)=0$$

This means that either $X=0$ or $(D-\mu)=0$, so a non-zero cell population can only be maintained when $D=\mu$.

In other words, a non-zero cell population will be maintained if the specific growth rate μ is balanced by the dilution rate D.

Once the organism has adjusted its specific growth rate to equal the dilution rate, the expression

$$DX_0=X(D-\mu)$$

can be satisfied by any value of X greater than zero, and this has been confirmed experimentally as shown in Fig. 3.4.[6] The above situation holds true only when:

- the specific growth rate is independent of the cell population
- growth is in the exponential phase and not affected by a decreasing concentration of a limiting nutrient.

MONOD CHEMOSTAT MODEL[2]

If, as in the case of batch culture, one of the nutrients is at growth-limiting concentrations, we can rewrite the cell balance at a steady state using the Monod expression for specific growth rate μ:

$$DX_0=X(D-\mu)=X\left(D-\frac{\mu_m C_i}{C_i+K_i}\right) \tag{3.5}$$

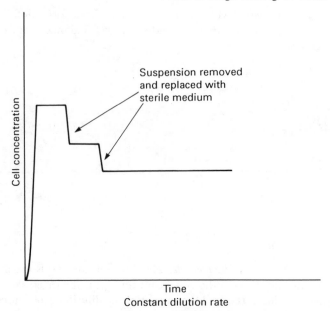

Time
Constant dilution rate

Fig. 3.4 Typical cell concentrations in continuous culture.

and if we define a yield factor Y as

$$Y = \frac{\text{mass of cells formed}}{\text{nutrient consumed}}$$

a steady-state balance on the limiting nutrient becomes

$$D(C_0 - C_i) - \frac{r}{Y} = 0$$

where C_0 is the initial concentration and C_i is the equilibrium nutrient concentration at any time. But

$$r = \mu X \text{ and } \mu = \frac{\mu_m C_i}{(C_i + K_i)}$$

therefore

$$D(C_0 - C_i) - \frac{\mu_m C_i X}{Y(C_i + K_i)} = 0 \qquad\qquad (3.6)$$

Equations (3.5) and (3.6) are usually called the *Monod chemostat model.*

For the case where $X_0 = 0$ (sterile feed), (3.5) and (3.6) can be solved for $(C_i)_{\text{SF}}$ and X_{SF}:

$$(C_i)_{SF} = \frac{DK_i}{(\mu_m - D)} \tag{3.7}$$

$$X_{SF} = Y(C_0 - C_i) \tag{3.8}$$

If the value of D is small $(D \to 0)$

$(C_i)_{SF} \to 0$ (cells consume the nutrient)

and

$$X_{SF} \to C_0$$

As

$$D \to \mu_m, \qquad X_{SF} \to 0$$

Once the dilution rate has exceeded the maximum possible growth rate, the only solution to (3.8) is $X = 0$.

This situation, where D exceeds D_{max} and all cells are lost from the system, is known as *washout*. When $X = 0$ in (3.8), $C_i = C_0$, and

$$C_0 = \frac{DK_i}{\mu_m - D}$$

thus

$$D_{max} = \frac{\mu_m C_0}{C_0 + K_i}$$

The pattern of behaviour of a single CSTF is shown in Fig. 3.5.

Near washout, the fermentor is very sensitive to variations in dilution rate (D), a small change in D affecting X and C_i disproportionately.

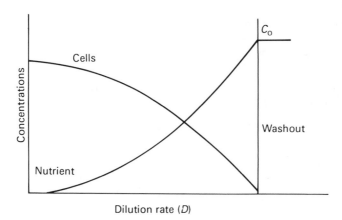

Fig. 3.5   Behaviour pattern in a single CSTF.

Table 3.1   Summary of notation used in Chapter 3

A	concentration of limiting nutrient A	kg/m^3
B	constant in $\mu = \mu_m C_i/(C_i + BN)$	kg/m^3
C	limiting nutrient concentration	kg/m^3
c	concentration in cell of essential component	
D	dilution rate	s^{-1}
F	volumetric flowrate	m^3/s
K	rate of increase of mould pellet	m/s
K_A	proportionality constant for utilization of A	$(kg/s)/no; s^{-1}$
K_i	saturation constant	kg/m^3
M	mass of mould pellet	kg
N	number of cells	no/m^3
R	radius of mould pellet	m
r	rate of cell formation ($= dX/d\theta$)	$(kg/m^3)/s$
V	vessel (fermentor) volume	m^3
X	cell concentration	$no/m^3; kg/m^3$
X_0	initial cell concentration	$no/m^3; kg/m^3$
X_S	static (maximum) cell concentration	$no/m^3; kg/m^3$
Y	yield factor	kg/kg nutrient
μ	specific growth rate	s^{-1}
μ_m	maximum specific growth rate	s^{-1}
ρ	density of mould pellet	kg/m^3
θ	time	s
θ_d	mass doubling time	$s; min$

subscripts

i	refers to nutrient i
p	refers to product conditions
sf	sterile feed conditions

Optimum Conditions

The rate of cell production per unit volume of fermentor is DX (FX/V), and the maximum cell production rate will be given by

$$\frac{d(DX)}{dD} = 0$$

(differentiating and equating to zero will give an optimum).

If we apply this to (3.7) and (3.8),

$$\frac{d}{dD} DY \left(C_0 - \frac{DK_i}{(\mu_m - D)} \right)$$

and for optimum cell production

$$D_{opt} = \mu_m - \left[\frac{K_i}{(C_0 + K_i)} \right]^{\frac{1}{2}}$$

Thus, if $C_0 \gg K_i$ (which is usually the case), D_{opt} approaches μ_m near washout.

In terms of control of the CSTF, this means that because X and C_i are sensitive near washout, we normally do not aim for the optimum (maximum) cell production, otherwise the fermentor operation may become unstable. In practice, operation is carried out at a dilution rate of about 80 per cent of D_{max}.

A summary of notation used in this chapter is shown in Table 3.1.

References

1. J. Monod, *Recherches sur des Croissances des Cultures Bacteriennes*, 2nd edn, Hermann et Cie, Paris (1942).
2. J.E. Bailey and D.F. Ollis, *Biochemical Engineering Fundamentals*, 2nd edn, McGraw-Hill, New York (1986).
3. S. Aiba, M. Shoda, and M. Nagatani, 'Kinetics of product inhibition in alcohol fermentation', *Biotech. Bioeng.* **10**, 845 (1968).
4. B. Metz and N.W.F. Kossen, 'The growth of moulds in the form of pellets—A literature review', *Biotech. Bioeng.* **19**, 781 (1977).
5. J.M. Coulson, J.F. Richardson, and D.G. Peacock, *Chemical Engineering*, Vol. 3, 2nd edn, Pergamon Press, Oxford (1979).
6. R.K. Finn and R.E. Wilson, 'Population dynamics of a continuous propagator for micro-organisms', *J. Agric. Fd Chem.* **2**, 66 (1954).

Chapter 4

Fermentor Design

The previous chapter dealt with the design of a fermentor as a device for cell culture and product generation (bioreactor), and was concerned with the prediction of the cycle time for batch operation or 'residence' time for continuous operation. This information, together with the available time for processing, allows us to obtain the physical size of the fermentor in terms of its volumetric capacity.

However, the fermentor is really a multi-purpose processor, and in practice has to perform a number of functions:

- as a *bioreactor* it must be sized to provide the required production capacity
- as a *piece of mass transfer equipment* it must be designed to ensure that nutrients and cells are well dispersed, and for aerobic processes that an adequate dissolved oxygen concentration is maintained and made readily available to the cells
- as a *control device* it must not only provide for the control of temperature, but also for the rapid dispersion of control chemicals (for pH and foaming), and to ensure that any sampling will give a representative picture of the important parameters (cell population, pH, oxygen concentration)
- as a *heat transfer device* it must ensure that a constant temperature is maintained during the growth cycle by the removal of heat generated from a variety of sources, and in certain circumstances it must be capable of sterilizing the medium *in situ*.

Thus, having first obtained the required volume of the fermentor, consideration must be given to the inclusion of the other functions in the ultimate fermentor specification.

This chapter is mainly concerned with the design of the fermentor as an efficient mixing and aeration device. Chapter 5 will consider the heat transfer aspects of the fermentor.

Aeration

In general, the transport mechanisms involved in the growth of a micro-organism can be shown as in Fig. 4.1. Figure 4.1 is a simple model, and in the case of bacteria, the major problem is to get the oxygen to the organism in sufficient quantities to enable exponential growth to occur.

The resistances to mass transfer from the medium to the organism can be listed as:

- diffusion from the bulk gas to the gas/liquid interface
- solution of the gas in the liquid at the interface
- diffusion of the dissolved gas into the bulk of the liquid medium
- transport of the dissolved gas to the immediate region of the organism
- diffusion through the stagnant region of liquid surrounding the organism
- diffusion into the cell
- consumption by the organism depending on the growth kinetics.

BASIC MASS TRANSFER

The rate of mass transfer of a gas into a liquid can be represented by the basic relationship

rate = driving force/resistance

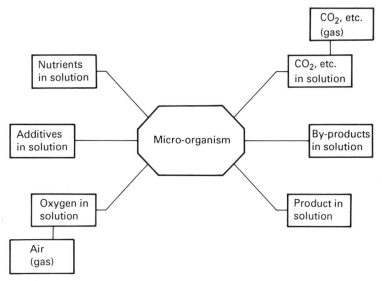

Fig. 4.1 Transport mechanisms in aerobic growth.

which is analogous to Ohm's law in electricity

current = p.d./resistance

This basic mass transfer relationship can be written as:

$$N_A = k_G(C_1 - C_2) = k_L(C_3 - C_4)$$

Here N_A is the rate of mass transfer of gas A ($kmol/s\, m^2$), C_1 and C_2 are gas concentrations in the gas phase, C_3 and C_4 are gas concentrations in the liquid phase, and k_G and k_L are gas and liquid mass transfer coefficients [with dimension resistance^{-1}].

The simplest theory of mass transfer is the *twin film theory*[1] and the conditions at the gas/liquid interface can be represented as shown in Fig. 4.2, by a plot of concentrations against distance from the interface. The model shown in Fig. 4.2 enables the simple rate law to be written as:

$$N_A = k_G(p_A - p_I) = k_L(C_I - C_A)$$

where p is the partial pressure of gas in the gas mixture (e.g. oxygen in air); C is the concentration of gas dissolved in the liquid; the subscript I refers to interface conditions; the subscript A refers to bulk conditions.

The regions near to the interface bounded by the interface and where the bulk concentrations begin to change are known as the *boundary layers*.

The mass transfer rate equations

$$N_A = k_G(\Delta p) = k_L(\Delta C)$$

assume a straight-line relationship over the boundary layers.

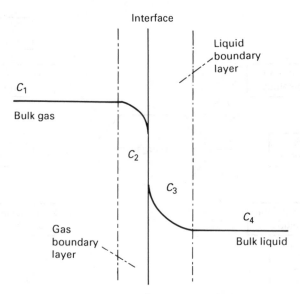

Fig. 4.2   Concentrations at a gas/liquid interface.

The mass transfer coefficients (k_G, k_L) are, strictly speaking, *film mass transfer coefficients* related to conditions and concentrations *at the interface*.

Because it is difficult to measure values of p_1 and C_1 we make use of *overall mass transfer coefficients* related to concentrations and conditions in the *bulk of the gas and liquid*. Overall coefficients are defined by

$$N_A = K_G(p_A - p^*) = K_L(C^* - C_A)$$

where p^* is the partial pressure of the gas in equilibrium with a solution of the gas of concentration C_A and C^* is the equilibrium concentration of the gas in solution giving a partial pressure of p_A above the solution.

For a large number of gases soluble in water (or fermentor medium), *Henry's Law* applies to equilibrium data:

$$p^* = HC_A$$

where H is Henry's constant.

If Henry's Law applies, then the following relationships are valid:

$$p_A = HC^*; \qquad p^* = HC_A; \qquad p_1 = HC_1$$

These relationships for equilibrium conditions can be used to derive the interrelationships between *film* coefficients and *overall* coefficients, and

$$\frac{1}{K_L} = \frac{1}{k_L} + \frac{1}{Hk_G}$$

$$\frac{1}{K_G} = \frac{1}{k_G} + \frac{H}{k_L}$$

For sparingly soluble gases like oxygen (10 ppm at 1 atmosphere pressure), the numerical value of H is very large (at $30\,^\circ$C, $H = 4.85 \times 10^4$ atmospheres/mol fraction), and

$$\frac{1}{K_L} \simeq \frac{1}{k_L}$$

In other words, the major resistance to mass transfer lies in the liquid film (boundary layer).

Because the rate of mass transfer (N_A) is the rate per unit area perpendicular to the direction of transfer, the total flow or transfer of gas is given by

$$K_L a(C^* - C_A)$$

where a is the transfer surface area per unit volume of medium.

In order to provide a large surface area for mass transfer, either

- the gas is broken up into small bubbles, or
- the medium (substrate) must be spread out over a large surface area.

The second option above is unsuitable for aseptic operation, and the main aeration technique used in fermentation where sterility is important is bubble formation.

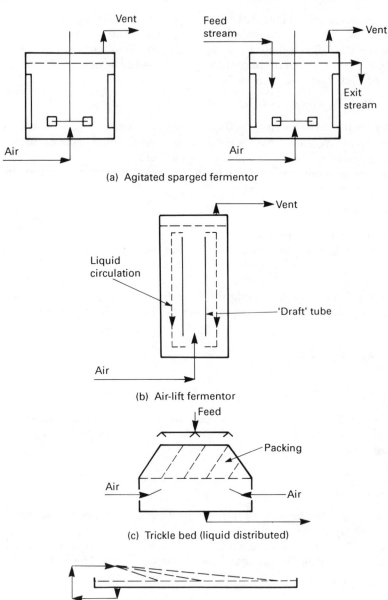

(a) Agitated sparged fermentor

(b) Air-lift fermentor

(c) Trickle bed (liquid distributed)

(d) Spray pond (liquid distributed)

Fig. 4.3   Fermentor types in common use.

Figure 4.3 shows the typical fermentor types which can be used to carry out aerobic processes. The most common device used is the mechanically agitated vessel using a flat-blade turbine agitator with a circular sparge ring fitted underneath the impeller to admit the air.

Figure 4.4 shows typical dimension ratios of an agitated fermentor, and Fig. 4.5 typical dimension ratios of the flat-blade turbine and sparge ring used in such vessels.

The reason for expressing the dimensions in terms of ratios is that, provided these ratios are maintained (independent of the physical size of fermentor) the 'scale-up' of the fermentor in terms of power requirements, aeration, agitator speed, etc. is simplified, particularly if 'geometrical' and 'dynamic' similarity are maintained (see pp. 53–5).

The agitation keeps the bubbles of air circulating through the medium. The faster the speed of rotation the longer the bubbles take to pass through the liquid in the fermentor before exiting, and the more time is therefore available for mass transfer.

The overall mass transfer coefficient (K_L) for such vessels has been shown by a number of workers[2, 3] to be related to the impeller diameter D_i and speed of revolution N by

$$K_L D_i \propto (N D_i^2)^{0.5}$$

and for constant design of impeller:

$$K_L \propto N^{0.5}$$

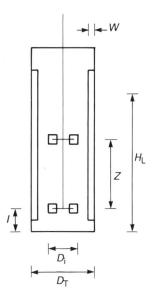

H_L/D_T	D_i	l	W	Z
1	$D_T/3$	D_i	$0.1\,D_T$	N/A
2	$D_T/3$	D_i	$0.1\,D_T$	D_T
3	$D_T/3$	D_i	$0.1\,D_T$	D_T

Fig. 4.4 Typical dimensions of an agitated fermentor.

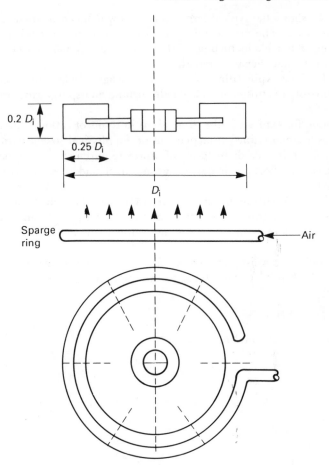

Fig. 4.5   Flat-blade turbine and sparge ring.

The work on oxygen absorption into fermentor media has been correlated by:[4]

$$K_L a = K_V = 0.002 \left(\frac{\Pi_G}{V} \right)^{0.7} V_S^{0.2} \qquad (4.1)$$

where K_V is the overall volumetric oxygen transfer coefficient $(1/s)$; Π_G/V is the power input per unit volume of liquid in a 'gassed' system (W/m^3), and V_S is the superficial air velocity based on the empty cross-sectional area of the vessel (m/s).

This expression applies for a liquid height (L) to vessel diameter (D_T) ratio of 1.0. For L/D_T ratios between 2 and 4, values of K_V are approximately 50 per cent greater than those given by the above correlation.

The total transfer rate is equal to $K_V(C^* - C_A)$ and the units of C will determine the units of K_V; for example, if C is measured in $kmol/m^3$ then K_V will be in $kmol/s\ m^3$ volume.

There are two important criteria for the maintenance of cell growth in aerobic fermentations:

- the maintenance of a dissolved oxygen concentration which is above the critical concentration required for exponential growth of the organism, similar to the required concentration of an essential nutrient (typical values ranging from 0.003 to $0.05\ mol/m^3$)
- the supply of oxygen at a rate matching the metabolic utilization by the organism.

Example 4.1

Experimental work on the growth of a micro-organism shows that the critical dissolved oxygen concentration required to sustain growth is 0.022 mmol/L, and the oxygen utilization rate is $1.68 \times 10^{-4}\ kmol/s\ m^3$.

Determine the required mass transfer coefficient for this process. At 30 °C, Henry's constant for oxygen/broth is 4.75×10^4 atmospheres/mol fraction.

Answer

The oxygen transfer rate is

$$K_La(C^* - C) = 1.68 \times 10^{-4}\ kmol/s\ m^3$$

Allowing for a dissolved oxygen concentration in the broth 20 per cent higher than the critical,

$$C = 1.2 \times 0.022 \times 10^{-3}\ kmol/m^3$$
$$= 2.64 \times 10^{-5}$$

At 1 atmosphere pressure, the partial pressure of oxygen in the air, assuming ideal gas behaviour, will be 0.21 atmospheres (21 per cent v/v).

Thus

$$x^* = \frac{p}{H} = \frac{0.21}{4.75 \times 10^4}\ \text{mol fraction}$$

But $x^*/(1 - x^*)$ is the molar ratio of oxygen/substrate. If ρ is the substrate density and M is the molecular weight of substrate, then

$$C^* = \frac{x^*\rho}{(M[1 - x^*])}\ kmol/m^3$$

Assuming $\rho = 1050\ kg/m^3$ and $M = 20\ kg/kmol$,

$$C^* = \frac{0.21 \times 1050}{4.75 \times 10^4 \times 20}$$
$$= 2.32 \times 10^{-4}\ kmol/m^3$$

since $(1 - x^*) \approx 1.0$. Thus, utilization rate

$$1.68 \times 10^{-4} = K_La(23.2 - 2.64) \times 10^{-5}$$

and

$$K_L a = 0.817 \, \text{kmol/s} \, \text{m}^3$$

Since

$$K_L a = 0.002 \left(\frac{\Pi_G}{V} \right)^{0.7} V_s^{0.2}$$

(from Equation 4.1), (Π_G/V) can be calculated for any particular fermentor.

Mixing and Agitation

The only difference between aerobic and anaerobic operation is that no air supply is required for anaerobic operation.

The need for agitation of the fermentor broth applies equally to both types of operation, in order to provide the dispersion functions described at the beginning of this chapter.

When a liquid is placed in a cylindrical vessel and a centrally mounted agitator is revolved, if no baffles are fitted, the main fluid motion is circumferential and the overall mixing is poor. Baffles increase the speed of mixing but at the expense of higher power requirements (see Fig. 4.6).

In order to design an agitated frementor, it is important to be able to predict the power requirements, particularly for aerobic fermentations, since the oxygen transfer rate is a function of the power input per volume (Π_G/V—Equation 4.1).

Practical problems in the applied sciences and engineering fall into two categories:

- problems where the behaviour of the system and materials are well defined in mathematical terms
- problems where the behaviour of the system and materials are either only partly defined, or are interdependent and cannot be defined in precise mathematical terms.

The first type of problem can be solved by strict mathematical methods, although the methods may require complicated procedures such as the integration of simultaneous differential equations. An example of this type of problem is in liquid sterilization (Chapter 2), where the thermal death kinetics are well defined so that the rate of reaction is

$$\frac{d\mathcal{N}}{d\theta} = -k\mathcal{N}$$

and the property $k = A \exp(-E/RT)$.

It is also possible to express the temperature profiles for heating and cooling

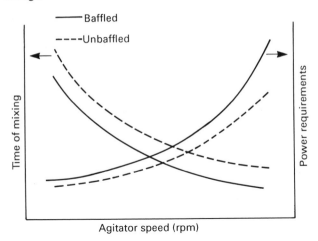

Fig. 4.6 Agitation with and without baffles.

mathematically (see Chapter 5), and the combination of these equations would enable a mathematical solution to be obtained for the value of ∇.

Problems of the second type above where no strict mathematical solution is possible can however be solved empirically using the technique of *dimensional analysis*.[5]

DIMENSIONAL ANALYSIS

In order to be able to use dimensional analysis, enough information of the physics of the system must be known, otherwise misleading results can arise if an important variable is omitted.

The technique is carried out by initially writing down the possible behavioural relationship of the variable we are interested in as follows:

Y = function (other variables; physical properties, etc.)
   $= K(A^a B^b C^c \ldots \text{etc.})$

where K is a dimensionless constant.

The next step is then to express the variables and physical properties in terms of basic dimensions, for example mass, length, time, etc.

So, for instance, velocity (m/s) can be expressed dimensionally as length/time, acceleration (m/s^2) as length/time^2.

Having rewritten our basic expression in terms of dimensions, the exponents of the dimensions ($a, b, c, \ldots$ etc.) can be equated, and the simultaneous equations obtained can then be solved.

Example 4.2
The period of oscillation of a simple pendulum may be analysed dimensionally assuming that the period (P) is a function of the length of the pendulum (l), the acceleration due to

gravity (g), the mass of the pendulum (m) and the horizontal displacement from the balance point (x).

Determine the possible relationship between P and the other variables.

Answer

We can write the relationship between P and the other variables as

$$P = Kl^a x^b g^c m^d \qquad (4.2)$$

Using dimensions of length $[L]$, time $[\theta]$ and mass $[M]$

$$P = [\theta]; \qquad l = [L]; \qquad x = [L]; \qquad g = [L]/[\theta^2]; \qquad m = [M]$$

Thus, dimensionally:

$$[\theta] = [L]^a [L]^b [L]^c / [\theta]^{2c} [M]^d$$

Equating the exponents of the dimensions,

length $[L]$: $0 = a + b + c$
time $[\theta]$: $1 = -2c$,
mass $[M]$: $0 = d$

Because we have four unknowns but only three equations, one of the unknowns must be retained, and the other variables expressed in terms of the retained unknown. If b is retained,

$$d = 0, \qquad c = -\tfrac{1}{2}, \qquad a = (\tfrac{1}{2} - b)$$

We can now substitute back into our original equation (4.2) and

$$P = Kl^{(\frac{1}{2} - b)} x^b g^{-\frac{1}{2}} m^0$$

It can immediately be seen that the mass (m) of the pendulum does not affect the period (P), since $m^0 = 1$, thus

$$P = Kl^{(\frac{1}{2} - b)} x^b g^{-\frac{1}{2}}$$

Collecting terms with the same exponents together gives

$$P = K(x/l)^b (l/g)^{\frac{1}{2}}$$

and the final relationship can be written as

$$P(g/l)^{\frac{1}{2}} = K(x/l)^b$$

The group (x/l) is the ratio of two lengths and is *dimensionless*.

The group $P(g/l)^{\frac{1}{2}}$ is also dimensionless, $\{\text{time}(\text{length}/\text{length} \times \text{time}^2)^{\frac{1}{2}}\}$.

The analysis carried out in Example 4.2 above gives the *possible* relationship between the period of oscillation of the pendulum and the variables associated with the system.

Note that it does not give a *precise* result; we still do not know the value of the exponent b nor the constant K, and these values can be obtained only from experimental work on the system.

Experimental work does show, however, that when the displacement of a pendulum is small, the term $(x/l)^b$ becomes unity, and the value of the dimensionless constant K is 2π.

Thus, for all practical purposes,

$$P\left(\frac{g}{l}\right)^{\frac{1}{2}} = 2\pi$$

or

$$P = 2\pi\left(\frac{l}{g}\right)^{\frac{1}{2}}$$

Dimensionless groups are collections of physical properties and variables such that *irrespective* of the *units* used for the calculation, the value of the group (or number) will always be the same, provided *consistent* units are used, i.e. all SI, all engineering, all cgs, etc.

Chemical engineering problems associated with the unit operations of fluid flow, mass transfer, heat transfer, mixing etc., usually can only be solved using dimensional analysis, and some of the more common dimensionless groups (or numbers) are given in Table 4.1.

It should be noted that not all relationships quoted in the literature use dimensionless groups. An example of a *dimensional expression* is Equation (4.1), where the groups (Π_G/V) and (V_S) are *not* dimensionless. In such cases, it is *vitally important* that the quoted units are *strictly* adhered to, otherwise the wrong result will be obtained.

POWER REQUIREMENTS

If dimensional analysis is applied to agitation to obtain a relationship of the power required (Π) with variables and physical properties:

$$\Pi = KD_i^a N^b \rho^c \mu^d g^e$$

where D_i is the impeller diameter, N is the speed of rotation, ρ is the fluid density, μ is the fluid viscosity, and g is the acceleration due to gravity, then we obtain a relationship containing three dimensionless groups:

$$\frac{\Pi}{\rho N^3 D_i^5} = K\left(\frac{D_i^2 N\rho}{\mu}\right)^a \left(\frac{D_i N^2}{g}\right)^b$$

In this relationship:

$\Pi/(\rho N^3 D_i^5)$	Power number (Po)
$D_i^2 N\rho/\mu$	Reynolds number (Re)
$D_i N^2/g$	Froude number (Fr)

This relationship will apply to one particular agitator in a particular vessel with particular baffle configurations, but we can use the same relationship for different sizes of vessel and agitators provided that geometric and dynamic similarity exists.

For *geometric similarity*, the following ratios should be the same for both vessels:

D_T/D_i	vessel diameter/impeller diameter
L/D_i	liquid height/impeller diameter

Table 4.1   Dimensionless groups (numbers)

Name	*Group*	*Meaning*
Reynolds (Re)	$\dfrac{Dv\rho}{\mu}$ (fluid flow)	$\dfrac{\text{(inertia forces)}}{\text{(viscous forces)}}$
	$\dfrac{D^2 N\rho}{\mu}$ (agitation)	
Froude (Fr)	$\dfrac{DN^2}{g}$ (agitation)	$\dfrac{\text{(inertia forces)}}{\text{(gravity forces)}}$
	$\dfrac{(v)^2}{Lg}$	
Prandtl (Pr)	$\dfrac{C_p\mu}{k}$	$\dfrac{\text{(diffusion of momentum)}}{\text{(diffusion of heat)}}$
Schmidt (Sc)	$\dfrac{\mu}{\rho \mathscr{D}}$	$\dfrac{\text{(diffusion of momentum)}}{\text{(molecular diffusion)}}$
Nusselt (Nu)	$\dfrac{hD}{k}$	$\dfrac{\text{(gradient at boundary)}}{\text{(gradient thro' fluid)}}$
Peclet (Pe)	$\dfrac{Dv\rho C_p}{k}$ (Re. Pr)	$\dfrac{\text{(heat transfer by convection)}}{\text{(heat transfer by conduction)}}$
	$\dfrac{Dv}{\mathscr{D}}$ (Re. Sc)	$\dfrac{\text{(mass transfer by momentum)}}{\text{(mass transfer by diffusion)}}$
Weber (We)	$\dfrac{D^3 N^2\rho}{\sigma}$	$\dfrac{\text{(inertia forces)}}{\text{(rising bubble forces)}}$

D is the characteristic length (diameter of pipe, sweep diameter of agitator); v is fluid velocity; ρ is fluid density; μ is fluid viscosity; C_p is fluid specific heat capacity; k is fluid thermal conductivity; $\mathscr{D}$ is fluid diffusivity coefficient; σ is fluid surface tension; L is characteristic length dimension; N is speed of rotation.

I/D_i impeller height from base of vessel/impeller diameter
W/D_T baffle width/vessel diameter

For *dynamic similarity* two possible methods are:

- scale-up can be based on making sure that the Reynolds number (Re) is the same in both sizes of vessel. For constant physical properties this means that

$$(D_i^2 N)_{\text{vessel }1} = (D_i^2 N)_{\text{vessel }2}$$

This approach (constant Reynolds number) is suitable for anaerobic processes

- for aerobic processes since we are dealing with a two-phase system (liquid/gas bubbles), a better scale-up criterion is the Weber number (We) (see Table 4.1). For constant Weber number with constant physical properties

$$(D_i^3 N^2)_{\text{vessel }1} = (D_i^3 N^2)_{\text{vessel }2}$$

For the standard six-blade turbine, at Reynolds numbers greater than 5000, the value of the Power number (Po) becomes 5.8, and applies to a single turbine placed a distance D_i from the base of the vessel, i.e. $I/D_i = 1$.

The power characteristics for a six-blade turbine are shown in Fig. 4.7.

For multiple impellers[6]

$$\text{Po} = J(\text{Po})_1$$

where J is the number of impellers and $(\text{Po})_1$ is the power number for a single impeller.

Impeller dia = $D_t/3$: Baffle width = $D_t/10$

----- Unbaffled

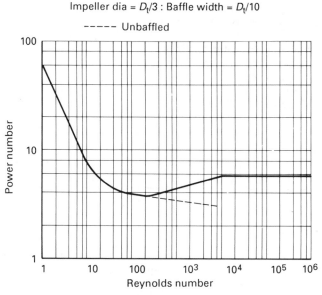

Fig. 4.7 Power curve for a six-blade turbine.

Table 4.2 Summary of notation used in Chapter 4

a	transfer surface of bubbles	m^2/m^3
C	concentration of gas in liquid	$kmol/m^3$
C_I	concentration at interface	$kmol/m^3$
D_i	impeller swept diameter	m
D_T	vessel diameter	m
$\mathcal{D}$	diffusivity coefficient	m^2/s
g	acceleration due to gravity	m/s^2
H	Henry's law constant	atmos/mol. fraction
I	impeller distance from vessel base	m
$\mathcal{J}$	number of impellers	—
K	dimensionless constant	—
K_G	overall gas mass transfer coefficient	$kmol/(s\,m^2\ atmosphere)$
K_L	overall liquid mass transfer coefficient	m/s
K_V	overall volumetric gas mass transfer coefficient	s^{-1}
k_G	individual gas mass transfer coefficient	$kmol/(s\,m^2\ atmosphere)$
k_L	individual liquid mass transfer coefficient	m/s
L	liquid height in vessel	m
l	length of pendulum	m
m	mass of pendulum	kg
$\mathcal{N}$	speed of revolution	s^{-1}
$\mathcal{N}_A$	rate of transfer of gas A	$kmol/s\,m^2$
P	period of oscillation of pendulum	s
p	partial pressure of gas	atmos
Po	Power number	
Q	volumetric gas flowrate	m^3/s
v	fluid velocity	m/s
V	volumetric capacity of vessel	m^3
V_s	superficial gas velocity (empty vessel)	m/s
W	baffle width	m
x	horizontal displacement of pendulum	m
μ	fluid viscosity	$kg/m\,s$
Π	power input to impeller	W
ρ	fluid density	kg/m^3
σ	surface tension	N/m

subscripts
G   gas
L   liquid

superscript * refers to equilibrium conditions

AIR SPARGING

In aerobic operation, the power requirements are less if the system is 'gassed', that is if air is fed to the underside of the agitator. Michel and Miller[7] proposed a general correlation to cover this case (for a single impeller):

$$(Po)_G = 0.354 \left(\frac{Po^2 N D_i^3}{Q^{0.56}} \right)^{0.45} \tag{4.3}$$

where N is the impeller speed (rev/s), D_i is the impeller diameter (m), and Q is the volumetric air flowrate (m^3/s). The subscript G applies to the 'gassed' system.

This correlation was applied to large-scale fermentor experiments by Brown,[8] who also proposed the following alternative:

$$(Po)_G = 0.7 \exp(-0.9Q) \tag{4.4}$$

Equations (4.3) and (4.4) are dimensional, and only the specified units must be used.

Note: The power requirement per unit volume (Π_G/V) is required for calculation of the mass transfer coefficient (see p. 47).

In designing an aerobic fermentor, it is essential to calculate the 'ungassed' power requirement, otherwise the drive motor may be underspecified leading to motor burnout should the airflow be stopped or fail whilst the drive is running.

A summary of notation used in this chapter is shown in Table 4.2.

References

1. R.E. Treyball, *Mass Transfer Operations*, 3rd edn, McGraw-Hill, New York (1985).
2. J.H. Rushton, 'The use of pilot plant mixing data', *Chem. Eng. Prog.* **47**, 467 (1951).
3. J.W. Richards, 'Studies in aeration and agitation', *Progress in Ind. Microbiol.* **3**, 143 (1961).
4. K. Van't Riet, 'Review of measuring methods and results in non-viscous gas-liquid mass transfer in stirred vessels', *Ind. Eng. Chem. Process Des. Dev.* **18**, 357 (1979).
5. P.W. Bridgman, *Dimensional Analysis*, Yale University Press, Cambridge, MA (1946).
6. Y. Ohyama and K. Endoh, 'Power characteristics of gas-liquid contacting mixers', *Chem. Eng., Japan* **19**, 2 (1955).
7. B.J. Michel and S.A. Miller, 'Power requirements of gas-liquid agitated systems', *A.I.Ch.E J* **8**, 262 (1962).
8. D.E. Brown, 'Power requirements in a production-scale fermentor', in *Fluid Mixing*, p. N1, Inst. Chem. Engrs. Symp. Ser. No. 64, Rugby (1981).

Chapter 5

Basic Heat Transfer

Unlike most chemical processes, which often operate at high temperatures, biologically based processes take place at temperatures typically not far removed from ambient. However, heat transfer is an important aspect of several parts of a biotechnological process, and one must consider both the application and removal of heat from any given part.

Whether cooling or heating, such heat transfer may be a deliberate attempt to alter the temperature of an operation, or may be designed to restrain temperature changes generated by an operation.

Deliberate temperature alteration includes such operations as media steriliz-ation, product concentration by evaporation, or freeze drying. In such cases large amounts of heat may need to be transferred in relatively short times, and provision for this will have to be designed into the process plant.

Isothermal heat transfer is important in the control of a bioreactor's environ-mental temperature and in such processes as cell disruption (where large quantities of heat may be generated). Biocatalysts, be they viable cells or enzymes, usually have a well-defined temperature range over which they are optimally active. If the temperature is too low biological activity may be negligible, and a high temperature (applied for too long) may result in thermal degradation of the cells or enzyme. For most mesophilic organisms (that is organisms which normally grow at near ambient temperatures) and enzymes derived therefrom, this temperature range is below 40 °C. However, large-scale bioreactors often show a net heat generation because of, for example, frictional heating by rotating impellers, metabolic activity, or exothermic reactions. In these circumstances the reactor will need cooling. It should be noted that, in contrast to the needs of deliberate temperature alteration, the rate of heat transfer during these constant-temperature (isothermal) operations is usually low. Thus the size of the heating/cooling coil placed in a bioreactor will be governed by its predicted use; large for rapid heating, smaller for the maintenance of constant temperature.

The rest of this chapter is concerned with the nature and quantification of heat transfer, and how process design may influence or be influenced by the need to control temperature.

The basic mechanisms of heat transfer are:[1]

- *conduction:* the transfer of heat from one part of a solid to another part of the solid without appreciable movement of the particles of the solid
- *convection:* the transfer of heat from one part of a fluid to another part of the same fluid by means of the movement of 'parcels' of the fluid. There are two recognized mechanisms of convection:

  - *natural convection*, where the fluid movement is due to density differences caused by thermal gradients
  - *forced convection*, where the fluid motion is caused by an outside agency, e.g. a pump or agitator

- *Radiation:* the transfer of heat from one body to another body, physically separated, by means of wave motion through space.

Conduction

The temperature profile in a block of homogeneous solid material with heat applied to one face of the solid is shown in Fig. 5.1.

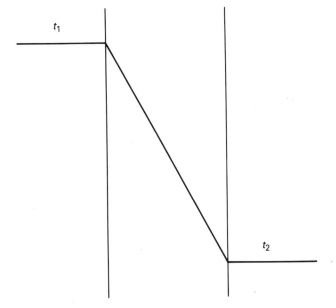

Fig. 5.1 Temperature profile through a solid block.

Under steady-state conditions (when the flow of heat and temperatures are constant with time at all points within the block), the rate of heat transfer per unit area (measured perpendicular to the flow of heat) Q/A, is given by

$$\frac{Q}{A} = \frac{k(t_1 - t_2)}{x} = \frac{k\Delta t}{x}$$

where x is the thickness of the block and Δt is the temperature difference between boundaries.

The proportionality constant (k) is known as the *thermal conductivity*, which is a specific property of the material of the block.

The above expression can also be written as

$$Q = \frac{\text{driving force}}{\text{resistance}}$$

$$= \frac{\Delta t}{R} \text{ where } R = \frac{x}{kA}$$

THREE MATERIALS IN CONTACT (SERIES)

The temperature profile at steady state in a composite block of three different materials in contact is shown in Fig. 5.2.

At a steady state, the rate of heat transfer through each part of the block will be the same, and therefore

$$Q_1 = Q_2 = Q_3 = Q$$

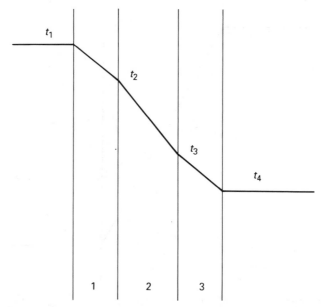

Fig. 5.2   Thermal gradients in a composite block.

and

$$Q = \frac{k_1 A (t_1 - t_2)}{x_1} = \frac{k_2 A (t_2 - t_3)}{x_2} = \frac{k_3 A (t_3 - t_4)}{x_3}$$

or

$$Q = \frac{\Delta t_1}{R_1} = \frac{\Delta t_2}{R_2} = \frac{\Delta t_3}{R_3} = \frac{\Delta t_0}{R_0}$$

where

$$R_1 = \frac{k_1}{x_1 A}; \qquad R_2 = \frac{k_2}{x_2 A}; \qquad R_3 = \frac{k_3}{x_3 A}$$

and

R_0 is the overall resistance.

This configuration is equivalent to resistances in series, and it can be shown that

$$R_0 = R_1 + R_2 + R_3$$

and

$$Q = \Delta t_0 / \left[\frac{x_1}{k_1 A} + \frac{x_2}{k_2 A} + \frac{x_3}{k_3 A} \right]$$

Example 5.1

A composite block of material made up from three different solids has the configuration shown in Fig. 5.3. Calculate the steady-state transfer of heat through the block given the properties shown.

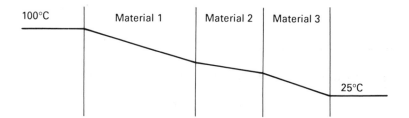

Material	Thermal conductivity W/m K	Thickness (mm)
1	28	100
2	65	60
3	40	60

Fig. 5.3 Composite block—Example 5.1.

Answer

Assuming that the cross-sectional area is $1\,m^2$,

$$R_1 = \frac{x_1}{k_1} = \frac{0.1}{28} = 0.00357$$

$$R_2 = \frac{x_2}{k_2} = \frac{0.06}{65} = 0.000923$$

$$R_3 = \frac{x_3}{k_3} = \frac{0.08}{40} = 0.002$$

and

$$R_0 = 0.00357 + 0.000923 + 0.002$$
$$= 0.006493$$
$$Q = \frac{\Delta t_0}{R_0} = \frac{(100 - 25)}{0.006493}$$
$$= 11\,551\ W/m^2$$

THICK-WALL CYLINDERS

Since the area required for heat transfer is measured at right angles to the direction of heat flow, certain configurations give rise to a continuously varying area as we move along the heat flow path.

Taking the example of a thick-wall tube heated from the inside surface with heat flowing to the colder outside of the tube, because the heat flow is radial, the area for heat transfer is a function of the particular radius at which it is measured.

In this particular case it can be shown that the mean area to be used is the *logarithmic mean area* (A_{LM}) given by:

$$A_{LM} = \frac{(A_0 - A_1)}{\ln(A_0/A_1)}$$

where A_0 is the outside surface area of the tube and A_1 is the inside surface area of the tube.

Example 5.2

(a) Calculate the log mean area of a tube of 50 mm inside diameter with a wall thickness of 25 mm.

(b) Calculate the arithmetic mean area of the same tube.

Answer

(a) Inside area/m length $= 0.05\pi$

Outside area/m length $= 0.1\pi$ (outside diameter $= 50 + 50$ mm)

$$A_{LM} = \frac{\pi(0.1 - 0.05)}{\ln(0.1/0.05)} = 0.227\ m^2/m\ \text{length}$$

(b) Arithmetic mean area $= \pi(0.1 + 0.05)/2 = 0.236\ m^2/m$ length.

Taking the arithmetic mean will give an error of 4 per cent on the area, and hence a 4 per cent error in the rate of heat transfer.

Commercial steel heat transfer tubes can be considered to be thin walled, and very little error is introduced by taking the arithmetic mean.

The error becomes significant in the case of lagging applied to pipes, where a thickness of 100–200 mm is not unusual.

Convection

If two fluids are flowing on either side of a solid wall (a situation which occurs in any heat exchanger), and if one of the fluids is at a higher temperature than the other, the temperature profiles near to the wall will be as shown in Fig. 5.4.

The two zones on either side of the wall where the temperature is varying are known as *thermal boundary layers* (to distinguish them from mass transfer boundary layers—see Chapter 4). The prediction of the temperature profiles and the thickness of the boundary layers is difficult because they are functions of the fluid velocity as well as the physical properties, which also vary with temperature.

If we assume that all the resistance to heat transfer through each fluid occurs in the boundary layer for that fluid, then the temperature profile can be taken as a straight line, and the situation becomes similar to that of conduction through a composite block of three materials (fluid 1, the solid wall, and fluid 2).

Thus, for this assumption, once a steady state has been reached:

$$Q = \frac{k_1 A \Delta t_1}{x_1} = \frac{k_w A \Delta t_w}{x_w} = \frac{k_2 A \Delta t_2}{x_2}$$

and

$$\Delta t_1 = (t_1 - t_2): \qquad \Delta t_w = (t_2 - t_3): \qquad \Delta t_2 = (t_3 - t_4)$$

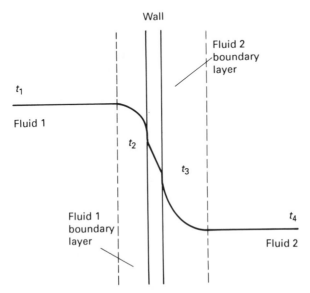

Fig. 5.4 Temperature profiles for convection between two fluids.

where x_1 and x_2 are the boundary layer thicknesses of fluids 1 and 2 respectively, and x_w is the wall thickness. Thus

$$Q = \Delta t_0 / \left(\frac{x_1}{k_1 A} + \frac{x_w}{k_w A} + \frac{x_2}{k_2 A} \right)$$

Unfortunately, the situation is more complex than for conduction because x_1 and x_2 are difficult to measure or to predict.

If, however, we define a *film heat transfer coefficient* (h) as

$$h = \frac{k}{x}$$

the above expression becomes

$$Q = \Delta t_0 A / \left(\frac{1}{h_1} + \frac{x_w}{k_w} + \frac{1}{h_2} \right)$$

where $h_1 = k_1 / x_1$ and $h_2 = k_2 / x_2$.

The thickness of the wall, x_w, is physically fixed, and the use of h_w is not necessary.

We can further define an *overall heat transfer coefficient* (U) as

$$\frac{1}{U} = \frac{1}{h_1} + \frac{x_w}{k_w} + \frac{1}{h_2}$$

and thus the rate of heat transfer (Q) becomes

$$Q = U A \Delta t_0$$

For any fluid, we can perform a dimensional analysis to obtain the likely relationship between the film heat transfer coefficient (h) and the system properties and characteristics:

- *natural convection:* since natural convection gives much lower rates of heat transfer than forced convection, almost all industrial heat exchangers operate under forced convection. The only exceptions are hot air radiators used for space heating, and the mechanism of heat transfer in this case is a combination of natural convection and radiation.
- *forced convection:* for the forced convection situation, dimensional analysis for the flow of a fluid inside a pipe heated on the outside gives the following relationship:

$$\frac{hD}{k} = K \left(\frac{DV\rho}{\mu} \right)^A \left(\frac{C_p \mu}{k} \right)^B$$

where D is pipe diameter, k is fluid thermal conductivity, V is fluid velocity, ρ is fluid density, μ is fluid viscosity, C_p is fluid specific heat capacity, and K is a constant.

The dimensionless groups (or numbers) are:

$$\frac{hD}{k} \quad \text{Nusselt number (Nu)}$$

$$\frac{DV\rho}{\mu} \quad \text{Reynolds number (Re)}$$

$$\frac{C_p\mu}{k} \quad \text{Prandtl number (Pr)}$$

The Reynolds number, which features in most fluid flow operations (see Chapter 4), is a criterion which describes the flow regime. Generally, if the value of the Reynolds number is less than 2100, *laminar* or *streamline* flow occurs. Above 2100, the flow is *turbulent*.[2]

The Prandtl number (Pr) contains only physical properties relating to the fluid.

For configurations other than pipes or tubes, the *equivalent diameter* (D_E) can be used in place of D:

$$D_E = (4 \times \text{cross-sectional area})/\text{wetted perimeter.}$$

Example 5.3
Calculate the equivalent diameter (D_E) for:

(a) a pipe diameter D
(b) an annulus with an inner diameter d and outer diameter D.

Answer
(a) cross-sectional area of pipe $= \pi D^2/4$
    wetted perimeter $= \pi D$
    equivalent diameter

$$D_E = \frac{4(\pi D^2/4)}{\pi D} = D$$

(b) cross-sectional area of annulus $= \pi(D^2 - d^2)/4$
    wetted perimeter $= \pi(D + d)$
    equivalent diameter $= (D - d)$

LAMINAR FLOW

For values of the Reynolds number (Re) less than 2100, the following relationship is used:[3]

$$\text{Nu} = 1.62\left(\text{Re. Pr.}\frac{D}{L}\right)^{0.33} \qquad L = \text{pipe or duct length}$$

Alternatively:

$$\text{Nu} = 2.06\left(\frac{WC_P}{kL}\right)^{0.33}$$

where WC_P/kL is the Graetz number and W is the mass flowrate of fluid.

All physical properties are evaluated at the *mean bulk temperature* (t_{BM}), where

$$t_{BM} = (t_{in} + t_{out})/2$$

TURBULENT FLOW

Provided that the Reynolds number $(Re) > 10\,000$, the Colburn[4] expression can be used:

$$Nu = 0.023(Re)^{0.8}(Pr)^{0.33}$$

The physical properties are evaluated at the mean bulk temperature, *except* viscosity, which must be evaluated at the *mean film temperature* (t_F), where

$$t_F = (t_{BM} + t_S)/2$$

and t_S is the surface temperature.

For the case of viscous oils where viscosity changes markedly with temperature, Sieders and Tate[5] modified the above expressions to:

$$Nu = 1.86 \left(Re. \, Pr. \frac{D}{L} \right)^{0.33} \left(\frac{\mu}{\mu_S} \right)^{0.14} \qquad \text{for laminar flow}$$

$$Nu = 0.027(Re)^{0.8} \, (Pr)^{0.33} \left(\frac{\mu}{\mu_S} \right)^{0.14} \qquad \text{for turbulent flow}$$

where μ_S is the viscosity of the fluid at the surface temperature.

Agitated Vessels

There are two ways of transferring heat to and from agitated vessels:

- using jackets
- using internal coils

JACKETS

The heat transfer coefficient at the inside wall of the vessel (h_i) is given by:[6, 7]

$$\frac{h_i D_T}{k} = 0.36 \left(\frac{D_i^2 N \rho}{\mu} \right)^{0.67} (Pr)^{0.33} \left(\frac{\mu}{\mu_S} \right)^{0.14}$$

where D_T is the inside diameter of the vessel, D_i is the agitator diameter, N is the speed of revolution of the agitator, k is the thermal conductivity of the liquid and Pr is the Prandtl number for the liquid.

With jackets fitted to vessels, the surface area/unit volume falls as the diameter (and volume) of the vessel increases, and once a certain size has been exceeded, the area becomes too small to remove the heat of fermentation. Once this situation has occurred, the only alternative is to use internal coils.

Obviously the heat of fermentation will dictate to a certain extent the maximum size of vessel which will be suitable for jacketing, as shown in Table 5.1.[8]

Table 5.1 Maximum size of vessel suitable for jacket

	Heat of fermentation (*W/L*)	*Maximum size for jacket*
Alcohol	58	50 L
Citric acid	15	3 m³
Penicillin	11	6 m³

INTERNAL COILS

The heat transfer coefficient at the outside surface of a coil (h_0) fitted inside an agitated vessel is given by:[6, 7]

$$\frac{h_0 D_T}{k} = 0.90 \left(\frac{D_i^2 N \rho}{\mu} \right)^{0.67} (\mathrm{Pr})^{0.33} \left(\frac{\mu}{\mu_s} \right)^{0.14}$$

It should be noted that the vessel diameter (D_T) appears in the Nusselt number, *not* the coil diameter.

Both expressions (for jackets and coils) can be applied to both laminar and turbulent conditions, and are valid for Reynolds numbers over the whole range likely to be encountered ($500 < \mathrm{Re} < 500\,000$).

Temperature Differences

If we take a simple double pipe annular heat exchanger and use a hot liquid flowing in the outer pipe to heat a colder fluid flowing in the inner pipe, there are two ways to arrange the flow of the liquids:

- co-current or parallel (in the same direction)
- counter-current (in opposing directions)

These two arrangements are shown in Fig. 5.5, which also shows the temperature profiles for the same conditions of heating and cooling the two fluids.

An analysis of both cases can be carried out (heat balances based on both heat transfer characteristics and thermal transfer between the fluids) which shows that in the expression

$$Q = U A \Delta t$$

the correct temperature difference (Δt) to use is the *log mean temperature difference* (LMTD or Δt_{LM}).

The log mean temperature difference is defined by

$$\Delta t_{LM} = \frac{(\Delta T_1 - \Delta T_2)}{\ln (\Delta T_1 / \Delta T_2)}$$

ΔT_1 and ΔT_2 are the *terminal* temperature differences.

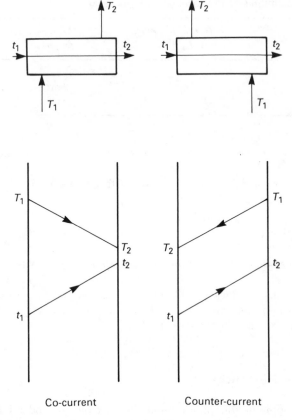

Fig. 5.5 Co-current and counter-current heat exchanger operation.

Example 5.4

Water is heated from 15 °C to 40 °C in a simple double-pipe heat exchanger using hot oil. The temperature of the oil entering the heat exchanger is 80 °C, and the oil exit temperature is 50 °C.

Calculate the log mean temperature difference for:

(a) co-current operation
(b) counter-current operation.

Answer

(a) *Co-current operation*

$$\Delta T_1 = (80 - 15) = 65$$
$$\Delta T_2 = (50 - 40) = 10$$
$$\Delta t_{LM} = \frac{(65 - 10)}{\ln (65/10)} = \mathbf{29.4}$$

(b) *Counter-current operation*

$$\Delta T_1 = (50 - 15) = 35$$
$$\Delta T_2 = (80 - 40) = 40$$

$$\Delta t_{LM} = \frac{(35 - 40)}{\ln (35/40)} = \textbf{37.4}$$

It can be seen from Example 5.4 that counter-current operation gives a higher temperature difference compared with co-current operation, and this offers a number of advantages:

- a higher value for Δt_{LM} will mean that the same duty can be carried out using a smaller heat exchanger—the area required (A) will be smaller if U is the same
- the temperature difference throughout the length of the heat exchanger does not show as large a variation as for the co-current case—as can be seen in Fig. 5.5
- the oil in Example 5.4 could have been allowed to leave the heat exchanger at a temperature less than 40 °C for counter-current operation (known as 'crossover').

In practical terms, wherever possible, heat exchangers are *always* operated counter-currently to take maximum advantage of the above points.

Fermentor Vessels

The case of heat transfer to and from a fermentor is slightly different from normal heat exchanger operation.

UNSTEADY-STATE HEATING AND COOLING

For heating a batch of liquid in a fermentor, the temperature of the contents is continually varying and hence the log mean temperature difference (Δt_{LM}) also varies. In the case of convection heat transfer with the contents initially at t_i and a heating fluid temperature t_S, it can be shown that[9]

$$\frac{(t - t_S)}{(t_i - t_S)} = \exp \left(-\frac{UA\theta}{mC_P} \right)$$

where t is the temperature at time θ, m is the mass of liquid in the vessel, C_P is the specific heat capacity of the liquid, and U is the overall heat transfer coefficient.

CONSTANT TEMPERATURE OPERATION

Where the fermentor contents are maintained at a constant temperature, the normal heat exchanger type of expression can be used. The log mean temperature

Table 5.2 Summary of notation used in Chapter 5

A	cross-sectional area for heat transfer	m^2
C_P	fluid specific heat capacity	$J/kg\,K$
D	diameter of pipe	m
D_E	equivalent diameter of duct or channel	m
D_i	diameter of agitator	m
D_T	inside vessel diameter	m
h	film heat transfer coefficient	$W/m^2\,K$
k	thermal conductivity of material	$W/m\,K$
L	length of pipe or duct	m
N	speed of revolution of agitator	s^{-1}
Q	rate of heat transfer	$W\,(J/s)$
R	resistance to heat transfer	K/W
R_0	overall resistance	K/W
ΔT	terminal temperature difference	K
t	temperature	C
U	overall heat transfer coefficient	$W/m^2\,K$
V	fluid velocity	m/s
W	mass flowrate of fluid	kg/s
x	thickness of material	m
μ	fluid viscosity	$kg/m\,s$
ρ	fluid density	kg/m^3

subscripts

LM	logarithmic mean
o	outside/overall
i	inside conditions
BM	mean bulk conditions
F	mean film conditions
s	heated surface conditions

difference (Δt_{LM}) is calculated using terminal temperature differences, but the vessel contents are at constant temperature.

Unlike counter-current operation, the exit temperature of the heating/cooling fluid in the jacket or coil cannot 'crossover', and for economic reasons, the maximum outlet temperature of fluid in the jacket or coil should not approach nearer than 5 °C to the temperature of the contents.

A summary of notation used in this chapter is shown in Table 5.2.

References

1. J.M. Coulson, J.F. Richardson, J.R. Backhurst, and J.H. Harker, *Chemical Engineering*, Vol. 1, 3rd edn, Pergamon Press, Oxford (1978).
2. O. Reynolds, 'On the dynamical theory of incompressible viscous fluids and the determination of the criteria', *Phil. Trans. R. Soc.* **177**, 157 (1886).

3. T.B. Drew, H.C. Hottel, and W.H. McAdams, 'Heat transmission', *Trans. A.I.Ch.E.* **32**, 271 (1936).

4. A.P. Colburn, 'A method of correlating forced convection heat transfer data and a comparison with fluid friction', *Trans. A.I.Ch.E.* **29**, 174 (1933).

5. E.N. Sieder and G.E. Tate, 'Heat transfer and pressure drop of liquids in tubes', *Ind. Eng. Chem.* **28**, 1429 (1936).

6. T.H. Chilton, T.B. Drew, and R.H. Jebens, 'Heat transfer coefficients in agitated vessels', *Ind. Eng. Chem.* **36**, 510 (1944).

7. G.H. Cummings and A.S. West, 'Heat transfer data for kettles with jackets and coils', *Ind. Eng. Chem.* **42**, 2303 (1950).

8. A.T. Jackson, 'Some problems of industrial scale-up', *J. Biol. Ed.* **19**(1), 48 (1985).

9. A.T. Jackson and J. Lamb, *Calculations in Food and Chemical Engineering*, Macmillan Press, London (1981).

Chapter 6

Separation Processes

Separation processes in chemical engineering are of two distinct types:

- *physical separations*

These are either based on the size of the materials or the difference in density of the materials, and apply to the separation of solid/liquid suspensions. Examples of this type of separation operation are:

- *filtration* of solid particles from a solid/liquid suspension using the particle size as a basis for separation
- *centrifugation* of a solid/liquid suspension, or of two immiscible liquids using the density difference between the materials
- *sedimentation and thickening* is the separation of a solid from a suspension based on density differences, but using gravitational forces to effect the separation.

- *diffusional separations*

This type of operation applies to the separation of the components of a single-phase, homogeneous mixture. The process is based on the ability (or otherwise) of the materials to form two phases, which may be solid, liquid or vapour, but the separations rely on the molecular diffusion properties of the materials. Examples of diffusional separation operations are:

- *distillation* of a solution of liquids by partial vaporization and re-condensation
- *evaporation* and removal of a liquid from a solid/liquid solution in order to increase the concentration of the solid in the solution
- *ultrafiltration and reverse osmosis* which are basically filtration processes using the difference in size of molecules to effectively concentrate the solution

- *absorption:* the separation of a gas mixture achieved by dissolving one of the gases in a liquid
- *adsorption:* separation of a gas mixture or liquid solution by using physical adsorption in a solid to preferentially 'trap' one of the gases or components of the liquid
- *liquid extraction* which involves the separation of a solution of liquids by the addition of another (essentially insoluble) liquid
- *drying* which refers to the removal of a liquid from within a solid mass by vaporization.

The remainder of this chapter is concerned with the physical separation operations of filtration and centrifugation. Concentration of solutions (diffusional operations) is dealt with in Chapter 7.

Filtration

Filtration applies to the separation of a solid suspension using a porous filter medium to retain the solid whilst allowing the liquid to pass through the bed of particles formed and the filter medium. The type of system is shown schematically in Fig. 6.1.

The theory of filtration is based on the flow of fluids through porous beds,[1, 2] but account must be taken of the fact that as filtration proceeds, the cake increases in thickness. The relationship between the flowrate of filtrate at any time and the bed properties is

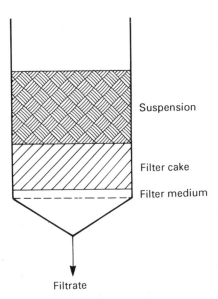

Filtrate

Fig. 6.1 Filtration of solid suspensions.

$$\frac{dV}{d\theta} = \frac{KA\Delta P}{\mu L} \tag{6.1}$$

where A is the cross-sectional area of the cake, L is the thickness of the cake, μ is the viscosity of the filtrate, ΔP is the pressure difference across the cake bed, V is volume of filtrate, θ is time and K is a constant for the particular bed of particles making up the cake.

It can be seen from equation (6.1) that the rate of filtration $(dV/d\theta)$ is proportional to the pressure difference (ΔP), and since large volumes of suspension must be handled in an industrial process, most filters operate under pressure to reduce the area (A) required for filtration (and to reduce the capital cost required).

The filter medium also offers a resistance to the flow of filtrate, and this is conveniently expressed as an *equivalent thickness* of filter cake (L_0). Thus the total thickness of cake becomes $(L + L_0)$.

Most filter cakes can be considered incompressible,[3] that is to say that properties of the cake such as K do not vary with applied pressure, and if we define a function

$$v = \frac{\text{volume of cake}}{\text{unit volume of filtrate}}$$

then a relationship between cake thickness (L), area (A), and filtrate volume (V) can be written:

$$LA = vV$$

Thus, Equation (6.1) above can be rewritten incorporating all the above assumptions as

$$\frac{dV}{d\theta} = \frac{KA\Delta P}{\mu(L + L_0)}$$

or

$$\frac{d\theta}{dV} = \frac{\mu L}{(KA\Delta P)} + \frac{\mu L_0}{(KA\Delta P)}$$

$$= \frac{\mu v V}{KA^2\Delta P} + \frac{\mu L_0}{KA\Delta P} \tag{6.2}$$

There are two ways in which a filter can be operated:

- *constant rate filtration*, where the pressure drop is continuously adjusted to keep the flowrate of filtrate $(dV/d\theta)$ constant
- *constant pressure filtration*, where the pressure drop remains constant (at pump pressure) and the rate of filtration decreases as the filter fills with cake.

The pattern of filtrate collection for both methods of operation is shown in Fig. 6.2.

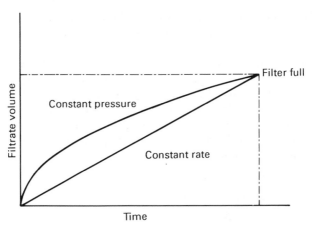

Fig. 6.2 Constant rate and constant pressure filtration.

CONSTANT RATE FILTRATION

Equation (6.2) above gives the differential inverse rate of filtration $(d\theta/dV)$ through a filter cake and filter medium at any time:

$$\frac{d\theta}{dV} = \frac{\mu v V}{KA^2\Delta P} + \frac{\mu L_0}{KA\Delta P}$$

For a constant rate of filtration

$$\frac{d\theta}{dV} = \frac{\theta}{V}$$

and the time of filtration required (θ) in order to collect a volume of filtrate (V) can be calculated if the other properties of the system are known:

$$\theta = \frac{\mu v V^2}{KA^2\Delta P} + \frac{\mu L_0 V}{KA\Delta P} \tag{6.3}$$

A plot of θ/V *vs.* V will be a straight line, and experimental evaluation of the value of K can be made together with an estimate of the equivalent cake thickness of the filter medium (L_0).

CONSTANT PRESSURE FILTRATION

Integration of the basic differential equation (6.2) is required because the rate of filtration is continually varying.

Integrating between $V=0$ at time 0 and $V=V$ at time θ gives

$$\theta = \frac{\mu v V^2}{2KA^2\Delta P} + \frac{\mu L_0 V}{KA\Delta P} \tag{6.4}$$

This expression assumes that constant pressure is achieved instantaneously at time 0. In practice, a finite time will be required to raise the system to operating pressure.

The normal industrial technique is to operate the filter under constant rate conditions by slowly raising the pressure until operating pressure is reached, and then to operate under constant pressure conditions.

This means that during the period of constant rate operation, a volume of filtrate V_0 is collected over a time period θ_0, and Equation (6.2) must be integrated from $\theta = \theta_0$ and $V = V_0$ for the constant pressure part of the cycle. This gives an integrated equation:

$$(\theta - \theta_0) = \frac{\mu v (V^2 - V_0^2)}{2KA^2 \Delta P} + \frac{\mu L_0 (V - V_0)}{KA \Delta P}$$

In this case a plot of $(\theta - \theta_0)/(V - V_0)$ *vs* $(V - V_0)$ will be a straight line because θ_0 and V_0 are constants once the constant-rate part of the operation has finished, and hence K and L_0 can be evaluated experimentally.

BATCH FILTRATION

Filtration as a batch operation consists of the following stages:

  (1) fill the filter with suspension and vent the air
  (2) filter
  (3) wash the cake (if required)
  (4) discharge the cake
  (5) re-assemble the filter ready for the next batch.

Descriptions and details of the construction of filters available for industrial use can be found in the literature,[4, 5] but the earlier type of 'plate and frame' filter press has been largely superseded by the *pressure leaf filter*. This type of filter consists of a pressure vessel inside which are located filter elements (leaves) of circular or rectangular construction. The filter medium is permanently fixed to the leaf by edge binding, and the leaf liquid outlets are connected individually to a common manifold which passes through the wall of the pressure vessel.

An example of a vertical leaf filter is shown in Fig. 6.3. During operation the suspension is fed to the vessel under pressure and solids build up on the leaf surfaces, the filtrate passing through the interior of the leaves to the outlet manifold.

The filter cake can be removed (after washing if required) as a solid through the large quick-opening port on the base, assisted by mechanical vibration of the leaves. As an alternative, the cake can be discharged as a slurry by washing the leaves with pressurized water jets.

An alternative leaf filter is shown in Fig. 6.4. The leaves, after filtration, are withdrawn from the pressure chamber to allow the filter cake to be readily discharged and collected.

This particular type of batch pressure filter finds many uses in the brewing, pharmaceutical, food processing and effluent treatment industries.

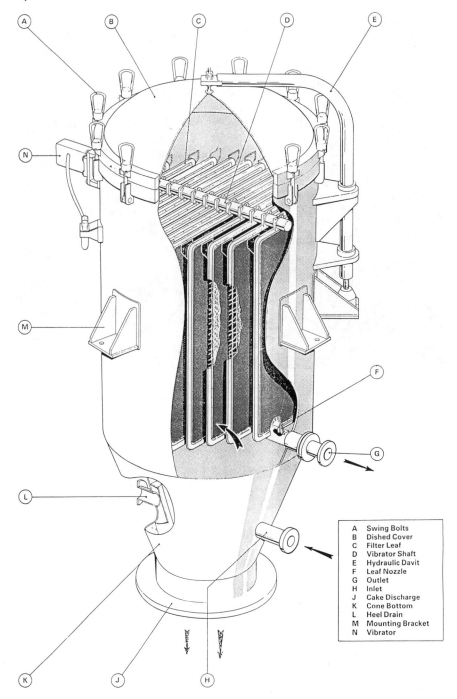

A	Swing Bolts
B	Dished Cover
C	Filter Leaf
D	Vibrator Shaft
E	Hydraulic Davit
F	Leaf Nozzle
G	Outlet
H	Inlet
J	Cake Discharge
K	Cone Bottom
L	Heel Drain
M	Mounting Bracket
N	Vibrator

Fig. 6.3 Vertical pressure leaf filter. (Courtesy of Stockdale Filtration Systems Limited.)

Fig. 6.4 Horizontal leaf filter. (Courtesy of Stockdale Filtration Systems Limited. Photograph by Brian Ollier.)

For constant pressure operation with an incompressible cake, time of filtration

$$\theta_F = aV^2 + bV$$

$$a = \frac{\mu v}{2A^2 \Delta P}; \qquad b = \frac{\mu L_0}{KA \Delta P}$$

If θ_W is the time required for washing/batch, and θ_A is the time for all operations other than filtering and washing, the total batch cycle time (Θ_T) will be

$$\Theta_T = \theta_F + \theta_W + \theta_A$$

If a volume V of filtrate is collected per batch, then the average output rate of filtrate (W) over the whole cycle of filtering, washing etc. is given by

$$W = \frac{V}{(\theta_F + \theta_W + \theta_A)}$$

The time of washing is a function of V, since the rate of washing the cake is proportional to the final rate of filtration, $dV/d\theta$.

Since $\theta_F = aV^2 + bV$, and $\theta_W = \beta V$ where β is a constant,

$$W = \frac{V}{(aV^2 + bV + \beta V + \theta_A)}$$

Differentiating with respect to V and equating to zero will give an optimum for V, and results in

$$V^2 = \frac{\theta_A}{a}$$

Using this result, the optimum time of filtering

$$\theta_F = aV^2 + bV$$

becomes

$$\theta_F = \theta_A + bV$$

and the time required for optimum filtration becomes the time required for the ancillary operations (all operations other than filtering and washing), plus an allowance for the resistance of the filter medium. Thus the size of filter required for optimum operation can be readily evaluated.[6]

CONTINUOUS FILTRATION

The most commonly used continuous filter is the *rotary drum filter*, usually operated under vacuum (Fig. 6.5). This type of filter is suitable for applications where large volumes of suspension need to be processed.

The filter medium is supported on a perforated drum which continually revolves in an agitated bath of suspension. Vacuum is applied to the inside of the drum, and the continuous flow of filtrate is removed through the vacuum system. The cake is carried round the drum circumference and can be removed continuously using a scraper knife or other means.

The rotary vacuum filter shown in Fig. 6.5 uses a system of cake discharge where the filter medium (cloth) leaves the drum at the discharge point and travels round a series of rollers, the sharp change in angle of the cloth causing the cake to come away from the cloth. This type of filter cake discharge allows the cloth to be cleaned (using water sprays) before it re-enters the filtering part of the cycle. An alternative cake discharge system is the knife discharge shown in Fig. 6.6.

This type of filter is used in the biotechnology industries for such operations as broth filtration, separation of precipitated solids and other large-scale suspension separations.

One advantage of continuous filtration using a rotary vacuum filter is that once the filter drum has completed the first revolution, operation takes place under constant pressure conditions.

Rather than use unit time as a design basis, it is more convenient to relate everything to a single revolution of the drum, and the time of filtration (depending

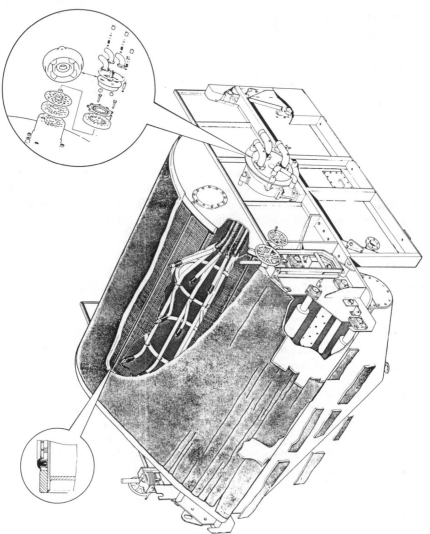

Fig. 6.5 Rotary vacuum filter. (Courtesy of Stockdale Filtration Systems Limited.)

Fig. 6.6 Knife discharge rotary vacuum filter. (Courtesy of Stockdale Filtration Systems Limited. Photograph by John Parkinson-Jones.)

on the submergence of the drum in the suspension) can be related to the other parameters required to apply Equation (6.4). The division of the various operations carried out during one revolution of the drum is shown in Fig. 6.7.

Centrifugation

A centrifuge consists basically of a basket in which a solid suspension (or two immiscible liquids) is rotated at high speed so that the two components are separated by centrifugal forces. The basket can be perforated or have solid walls, the perforated basket acting as a filter.

If a particle of mass m at a radius of r has an angular velocity of ω, then the centrifugal force acting on the particle will be $mr\omega^2$.

The ratio of the centrifugal force to the gravitational force will give a measure of the separating power of a particular centrifuge compared with gravity settling, and

$$Z = \frac{r\omega^2}{g}$$

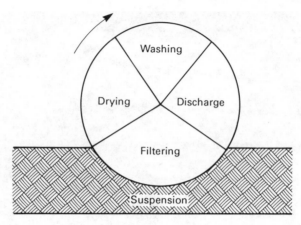

Fig. 6.7 Rotary drum filter operation.

Z often has a value of many thousands, and thus separation is much more efficient than gravity settling. Table 6.1 gives some comparative figures.

SEPARATION OF SOLIDS FROM LIQUIDS

The method used for the separation of a solid from a suspension in a centrifuge will depend on the characteristics of the bed of solid particles formed. Where the particles are small in size, and the bed formed would offer a high resistance to fluid flow, a solid wall basket is used, the solids being allowed to build up at the wall to a suitable depth before the centrifuge is stopped and the solids discharged. If the resistance to fluid flow of the particle bed is low, then a perforated basket can be used and the centrifuge operated as a filter (see Fig. 6.8).

The equation of motion of a particle in a liquid under centrifugal force is given by a simple force balance:

$$\begin{bmatrix} \text{radial} \\ \text{centrifugal} \\ \text{force} \end{bmatrix} - \begin{bmatrix} \text{drag force} \\ \text{opposing} \\ \text{motion} \end{bmatrix} - \begin{bmatrix} \text{accelerating} \\ \text{force on the} \\ \text{particle} \end{bmatrix} = 0$$

For a spherical particle in a centrifugal field,

$$\text{radial force} = \left(\frac{\pi d^3}{6}\right)(\rho_s - \rho)r\omega^2$$

where d is particle diameter, ρ_s is particle density, ρ is liquid density, ω is angular velocity, and r is the radius from the axis of revolution.

The drag force opposing particle motion will depend on the flow regime of the particle motion, and will be given by either[4]

$$3\pi\mu d\left(\frac{\partial r}{\partial \theta}\right) \quad \text{for laminar motion}$$

Table 6.1 Comparison between centrifugation and gravity settling

		Separation of sand $(\rho = 1300\,\text{kg/m}^3)$ in water Time for particle to move through 1 cm		
d	*Gravity*		*Speed of revolution*	
(μm)	*settling*	3000	10 000	30 000
50	10 min	1.38 min	8.3 s	—
10	4.3 h	38 min	3.4 min	23 s
5	11 h	2.6 h	13.8 min	1.5 min
1	18 days	—	6 h	0.6 h

or

$$0.22 \left(\frac{\partial r}{\partial \theta}\right)^2 \rho \left(\frac{\pi d^2}{4}\right) \quad \text{for turbulent motion}$$

The accelerating force on the particle is given by

$$\left(\frac{\pi d^3}{6}\right) \rho s \left(\frac{\partial^2 r}{\partial \theta^2}\right)$$

Laminar Conditions
The accelerating force on the particle (containing second-order differentials) is numerically small compared with the centrifugal and drag forces and can be neglected.

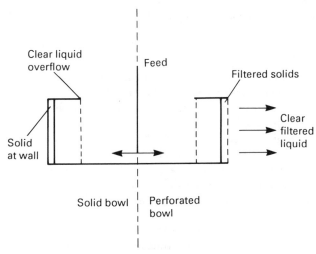

Fig. 6.8 Centrifuge operation for solid/liquid separation.

A modified force balance then becomes

$$\left(\frac{\pi d^3}{6}\right)(\rho_s-\rho)r\omega^2 = 3\pi\mu d\left(\frac{\partial r}{\partial \theta}\right)$$

Thus

$$\partial\theta = \left[\frac{18\mu}{(d^2\omega^2\Delta\rho)}\right]\left[\frac{\partial r}{r}\right]$$

where $\Delta\rho = (\rho_s-\rho)$.

The approximate time taken for a particle to travel from a radius of r_1 to a radius r_2 will be

$$\theta = \left[\frac{18\mu}{(d^2\omega^2\Delta\rho)}\right]\left[\ln\left(\frac{r_2}{r_1}\right)\right]$$

Turbulent Conditions

Neglecting the accelerating force, the force balance becomes

$$\left(\frac{\pi d^3}{6}\right)(\Delta\rho)r\omega^2 = 0.22\left(\frac{\partial r}{\partial\theta}\right)^2\rho\left(\frac{\pi d^2}{4}\right)$$

and

$$\left(\frac{\partial r}{\partial\theta}\right)^2 = 3d\omega^2 r\left(\frac{\Delta\rho}{\rho}\right)$$

or

$$\partial\theta = \frac{\partial r}{[3d\omega^2 r(\Delta\rho/\rho)]^{0.5}} = \left[3d\omega^2\left(\frac{\Delta\rho}{\rho}\right)\right]^{-0.5}\left(\frac{\partial r}{r}\right)^{0.5}$$

The approximate time taken for a particle to travel from a radius of r_1 to radius r_2 will be

$$\theta = 2[3d\omega^2(\Delta\rho/\rho)]^{-0.5}[r_2^{0.5} - r_1^{0.5}]$$

There will be a maximum rate at which a suspension can be fed to the centrifuge in order to obtain complete separation, since an adequate time must be allowed for each particle to travel from the inner liquid radius to the walls of the basket before the liquid leaves the machine.

If the inner liquid radius is r_1 and the bowl radius is r_2, then the maximum distance any particle will have to travel will be (r_2-r_1). The time taken for a particle to travel this distance can be calculated from the expressions given above, and must not exceed the time the particle remains in the centrifuge (as a function of the suspension flowrate). Thus the maximum throughput of the machine can be calculated.

If Q is the volumetric flowrate of suspension and V is the volume of liquid retained in the centrifuge, the average time any element of suspension remains in the centrifuge will be

$$\frac{V}{Q}$$

For laminar conditions

$$\theta = \left[\frac{18\mu}{(d^2\omega^2\Delta\rho)} \right] \ln\left(\frac{r_2}{r_1}\right)$$

where r_2 is the bowl radius and r_1 is the inner liquid radius.

If the flowrate is adjusted in order to just retain a particle of diameter d:

$$\theta = \frac{V}{Q}$$

and

$$Q = \frac{d^2\omega^2\Delta\rho V}{[18\mu \ln (r_2/r_1)]}$$

or

$$Q = \left[\frac{d^2\Delta\rho g}{18\mu}\right]\left[\frac{\omega^2 V}{g\ln(r_2/r_1)}\right]$$

$$= U_0\Sigma$$

u_0 is the free settling velocity of a particle diameter d under gravity, g is the acceleration due to gravity, and

$$\Sigma = \frac{\omega^2 V}{[g \ln (r_2/r_1)]}$$

Σ is a function only of the centrifuge size and its operation, and is a measure of the separation characteristics of the centrifuge.

It is common industrial practice to choose a particle diameter such that 50 per cent of that size will be collected, and 50 per cent lost in the liquid overflow (d_{50}). Particles larger than d_{50} will *all* be collected and smaller particles rejected.

Because particles are not spherical, and also because hindered settling occurs (particles interfere with each other), an efficiency term must be included in the expression for suspension flowrate. Thus

$$Q = 2u_0\eta\Sigma$$

and u_0 is calculated using d_{50}; η is the efficiency of the machine.

The value of Σ, together with the expected flowrate of the suspension, Q, can be used to select the particular type of centrifuge suitable for any operation.[6] Typical values are given in Table 6.2.

The tubular type of centrifuge is simply an empty, cylindrical bowl machine, usually of small diameter. As the size of this type of machine is increased, the distance a particle will have to travel from the inner liquid radius to the bowl wall increases considerably and the maximum possible flowrate of suspension decreases.

Table 6.2   Operating characteristics of centrifuges (based on $Q = 2u_0\eta\Sigma$)

Machine type	Flowrate range $Q(m^3/min)$	Q/Σ ($\times 10^8$ m/s)	Efficiency (η)
Tubular	0.01–0.1	5–35	0.98
Disc	0.002–2.0	8–50	0.5
Decanter	0.015–2.3	500–1.5×10^5	0.6

In order to reduce the distance a particle must travel before being collected at a solid surface, the disc machine was developed. A stack of conical discs placed in the bowl reduces the distance of travel to the horizontal distance between plates, and the collected particles travel along the underside of the plates to the bowl wall.

Figure 6.9 shows a disc machine with a hydraulically operated base section which allows intermittent discharge of the collected solids.

The decanter type of machine is a continuous centrifuge using a helical screw conveyor within the bowl to continuously remove the solids collected at the wall. This type of machine is shown in Fig. 6.10.

The various types of centrifuge available for industrial use are described in the literature.[4, 5, 7]

A summary of notation used in this chapter is shown in Table 6.3.

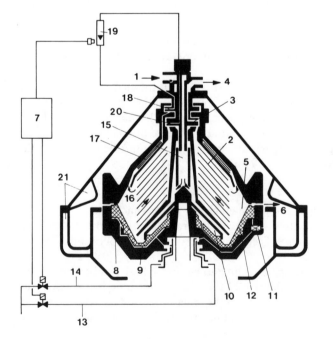

1 Feed
2 Discs
3 Centripetal pump
4 Discharge
5 Sediment holding space
6 Sediment ejection ports
7 Timing unit
8 Outer closing chamber
9 Inner closing chamber
10 Opening chamber
11 Bowl valve
12 Piston
13 Opening water
14 Closing water
15 Soft-stream inlet
16 Sensing liquid offtake
17 Clarifying discs for
     sensing liquid
18 Sensing liquid pump
19 Flowmeter
20 Sensing liquid pump
21 Cooling chambers

Fig. 6.9.   Typical disc centrifuge. (Courtesy of Westfalia Separators Limited.)

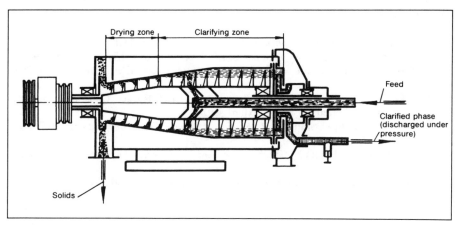

Fig. 6.10 Decanter continuous centrifuge. (Courtesy of Westfalia Separators Limited.)

Table 6.3 Summary of notation used in Chapter 6

A	cross-sectional area	m^2
d	diameter of particle	m
g	acceleration due to gravity	m/s^2
K	filter cake constant	m^2
L	thickness of filter cake	m
L_0	equivalent cake thickness of filter medium	m
m	mass of particle	kg
ΔP	pressure difference across filter (or membrane)	N/m^2
Q	suspension flowrate	m^3/s
r	radius from axis of rotation	m
V	volume of filtrate	m^3
	volumetric capacity of centrifuge	m^3
V_0	filtrate collected during constant rate period	m^3
v	volume of cake/unit volume of filtrate	m^3/m^3
W	average filtrate output rate	m^3/s
z	ratio of centrifugal/gravity forces	
β	permeability constant of filter cake	s/m^3
μ	fluid viscosity	$kg/m\ s$
ρ	fluid density	kg/m^3
ρ_s	solid density	kg/m^3
Σ	centrifuge separation criterion	—
θ	time	s
θ_0	time to collect V_0 filtrate	s
ω	angular velocity	$radians/s$

subscripts

A	ancillary operations
F	filtration
T	total batch cycle
W	washing

References

1. H.P.G. Darcy, *Les fontaines publique de la ville de Dijon*, Victor Dalamont, Paris (1856).
2. P.C. Carman, 'Flow through granular beds', *Trans. Inst. Chem. Engrs.* **15**, 150 (1937).
3. H.P. Grace, 'Resistance and compressibility of filter cakes', *Chem. Eng. Prog.* **49**, 303 (1953).
4. J.M. Coulson, J.F. Richardson, J.R. Backhurst and J.H. Harker, *Chemical Engineering*, Vol. 2, 3rd edn, Pergamon Press, Oxford (1978).
5. W.L. McCabe and J.C. Smith, *Unit Operations of Chemical Engineering*, 3rd edn, McGraw-Hill Kogakusha, Tokyo (1976).
6. A.T. Jackson and J. Lamb, *Calculations in Food and Chemical Engineering*, Macmillan, London (1981).
7. J.E. Bailey and D.F. Ollis, *Biochemical Engineering Fundamentals*, 2nd edn, McGraw-Hill, New York (1986).

Chapter 7

Concentration

Concentration of solutions produced by biological processes may be required for a variety of reasons:

- reduction of water content to reduce the costs of transport, e.g. concentration of beer wort produced at a central facility and shipped to other parts of the country
- reduction of water content because the product is marketed as a standardized, concentrated solution, e.g. enzyme solutions
- concentration of the solution prior to another process for reasons of economics, e.g. before drying.

A number of alternative processes are available for the concentration of solutions:

- thermal evaporation
- ultrafiltration and reverse osmosis
- freeze concentration
- adsorption/desorption processes

Evaporation[1]

Evaporation is the concentration of a water solution using thermal methods by vaporization of part of the water and subsequent separation from the solution. Evaporation is a suitable concentration method for:

- solid/liquid solutions
- solutions of two liquids where the solute liquid does not exhibit a large vapour pressure at the temperature of evaporation.

Table 7.1   Relationship between boiling point and temperature for water

Absolute pressure (bar)	*Boiling temperature* (°C)	*Vacuum* (ins Hg)
1.013	100	0.0
0.312	70	20.8
0.123	50	26.3
0.066	38	28.0
0.042	30	28.7

Table 7.2   Latent heat of vaporization of water

Latent heat (kJ/kg)	*Absolute pressure* (bar)	*Temperature* (°C)
2202	2.0	120.2
2258	1.0	100.0
2411	0.066	38.0

If the pressure of the system is reduced, the boiling point of the liquid will be reduced (the boiling point of a liquid being defined as the temperature at which it turns to vapour or the vapour pressure equals the system pressure).

Water has a boiling point/pressure relationship as shown in Table 7.1.

In order to ensure minimum thermal damage to heat sensitive material, it is normal practice to work under vacuum (economically at about 28 inches of mercury, giving a temperature of 38 °C).

Heat must be provided in order to vaporize the water, and steam is normally used. Approximately 1 kg of steam will evaporate 1 kg of water, but since the latent heat of vaporization of water at reduced pressures is higher than at elevated pressures (such as those at which steam is generated), slightly more steam will be required. Table 7.2 shows the variation of latent heat with pressure (and hence temperature).

Example 7.1
100 kg/h of a 5 per cent solution is to be concentrated to 25 per cent at 28 ins Hg vacuum using steam at 2.0 bar. This problem is shown diagrammatically in Fig. 7.1. Calculate the quantity of steam required to perform this operation.

Answer
Solute balance: $100 \times 0.05 = 0.25 \, Y$, thus $Y = 20$ kg/h
Overall balance: $100 = X + Y = X + 20$

and therefore $X = 80$ kg/h of water evaporated.

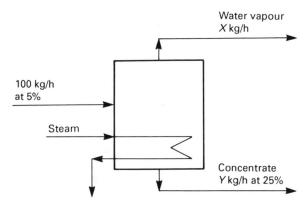

Fig. 7.1 Simple concentration—Example 7.1.

Steam requirement/kg water evaporated $= \dfrac{2411}{2202}$ kg (latent heat values taken from Table 7.2).

Thus, total steam requirement $= 80 \times \dfrac{2411}{2202}$

$$= 87.6 \text{ kg/h}$$

In an evaporation process, since the water evaporated is vapour (steam), although at a low pressure, it is worth trying to recover some of the thermal energy in the vapour in the interests of economy, and the simplest way is to use this vapour to heat another evaporator.

The use of *multiple-effect* evaporation is widespread throughout the process industries, and a typical triple effect arrangement is shown in Fig. 7.2.

Example 7.2
Taking the information in Example 7.1 but using two effects, recalculate the steam requirements.

Answer
The double effect arrangement for this problem is shown in Fig. 7.3.
Overall balance (both effects): $100 = S_1 + S_2 + Y$.
$Y = 20$ kg/h as before, but $(S_1 + S_2) = 80$ kg/h (total vaporization).
If we assume equal evaporation rates in both effects,

$$S_1 = S_2 = 40 \text{ kg/h in each effect.}$$

First effect: $100 = Z + S_1$ thus $Z = 60$ kg/h.

Steam use: steam is now only required to evaporate 40 kg/h in the first effect, and so the steam use will be 43.8 kg/h.

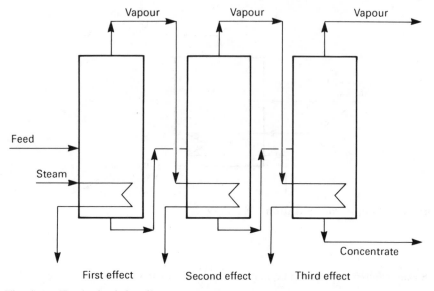

Fig. 7.2   Typical triple effect evaporator arrangement.

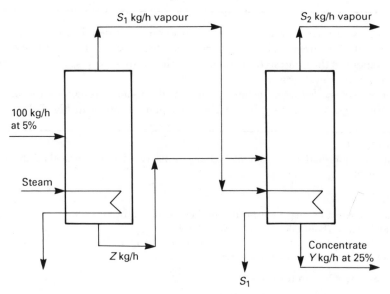

Fig. 7.3   Evaporator arrangement for Example 7.2.

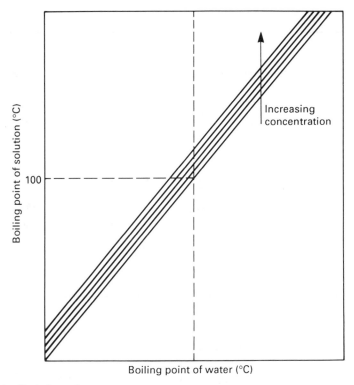

Fig. 7.4 Duhring plot.

Elevation of the Boiling Point
The addition of a solute to a pure liquid leads to a rise in the boiling point of the solution. This elevation of the boiling point is a function of concentration, e.g. 25 per cent caustic soda solution boils at 110 °C, 35 per cent at 125 °C.

There is no basic relationship which can be used to predict the elevation, and ·the best method to use for engineering design is a Duhring plot of boiling point of solution versus boiling point of water (see Fig. 7.4).

Although evaporation is a well-established process, and multiple effect operation is highly energy efficient, there are disadvantages:

- a high energy input in vaporizing water (latent heat of vaporization $= 2.3 \times 10^6$ J/kg)
- temperatures deleterious to many biological products occur (even though they are minimized by operation under vacuum), and for very heat-sensitive materials (e.g. antibiotics) evaporation is unsuitable.

Table 7.3 Typical membrane properties

Membrane	Pore size (μm)	Typical water flowrate (ml/s m²atmos)	Application	Maximum temperature (°C)
Asypor[7a]	0.2	260	UF	126
(asymetric	0.65	878	UF	
mixed cellulose				
esters)				
Nypor[7b]	0.2	115	UF	130
(nylon)	0.65	236	UF	
Tetpor[7c]	0.1	112	UF	140
(PTFE)	0.2	246	UF	
DuPont[2]	0.004	0.65	RO	90
250–PT–63				
(cellulose)				
Visking[2]	0.002	0.17	RO	80
(cellulose)				

Ultrafiltration and Reverse Osmosis[2, 3]

These two similar processes apply to the use of semipermeable membranes as molecular 'filters'. One of the differences between the two is that ultrafiltration tends to be a fractionation process (separating a range of molecular sizes), whereas reverse osmosis is applied to the separation of water molecules from a solution.

Although membrane separation has been known for a long time,[4] only the recent development of strong, anisotropic membranes for the desalination of seawater has made these processes competitive with other concentration processes.

Since the molecules are 'filtered' out of solution, the pore size of the membrane plays an important part in the efficiency of the process.

In terms of molecules, ultrafiltration (UF) is satisfactory for organic and biological molecules (molecular weight $\gg 300$). The size of these is such that they are retained on the surface of the membrane.

Reverse osmosis (RO) membranes are capable of retaining ions like Na^+ and Cl^-, but still allowing molecules of water to pass through.

Pore sizes of membranes used to carry out these two operations are:

- *ultrafiltration* $0.01–0.02\,\mu m$
- *reverse osmosis* $0.0001–0.01\,\mu m$

Table 7.3 shows some typical membrane characteristics.

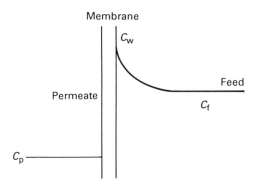

Fig. 7.5   Conditions in the region of the membrane.

ULTRAFILTRATION

Because the process is one of filtration, the concentration of the solute molecules being filtered out builds up at the surface of the membrane (known as concentration polarization). If the saturation concentration of the solute is moderately low, a gel layer can form on the membrane surface which may interfere with the permeation.

As in all filtration operations, the rate of filtrate (permeate) production is a function of pressure. The conditions in the region of the membrane are shown in Fig. 7.5.

Since the pores act as capillaries, the Hagen and Poiseuille[5] laminar flow equation can be used:

$$\phi = \frac{\Delta P}{32\mu R_M}$$

where ϕ is the rate of permeation, ΔP is the pressure drop across the membrane, μ is the permeate viscosity, and R_M is the flow resistance of the membrane.

If a gel layer forms due to polarization, the above equation must be modified to include a gel layer resistance, R_G, and

$$\phi = \frac{\Delta P}{32\mu(R_M + R_G)}$$

As a measure of the efficiency of the process, the *retention* (or *rejection*) *factor* (R) is defined by:

$$R = \frac{(C_F - C_P)}{C_F}$$

where C_F is the solute concentration in the feed and C_P is the solute concentration in the permeate. This retention factor (R) applies to both ultrafiltration and reverse osmosis.

In ultrafiltration, because of polarization, a steady state is reached when the diffusion of solute from the membrane back into the feed stream balances the rate of permeation, and:

$$\phi(C - C_P) = \mathscr{D}\left(\frac{dC}{dx}\right)$$

where $\mathscr{D}$ is the diffusion coefficient of solute in solution.

The pressures required in ultrafiltration are of the order of 7 atmospheres, and membranes are available in a wide range of polymeric materials (cellulosic, olefinic, aromatic, etc.).

REVERSE OSMOSIS

Reverse osmosis applies to the separation of smaller molecules than the ultrafiltration process, and also applies in the case of separation of salts (or solutions containing salts). Because of the presence of ionic salts, the physical phenomenon of osmosis plays an important part in the characterization of the process.

Osmotic Pressure
If a salt solution is separated from water by a semipermeable membrane, molecules of water will pass through the membrane in an attempt to dilute the salt solution and equalize the concentrations. In doing so the pressure on the salt solution side of the membrane will increase until the flow of water molecules is balanced by this pressure. Once this situation has occurred, this equilibrium pressure is called the *osmotic pressure*. The osmotic pressure increases as the salt concentration increases.

For concentration of a salt solution using an 'ultrafiltration' process, the water molecules must pass through the membrane in the opposite direction to that dictated by osmotic forces. In order to perform the operation, the osmotic pressure must be overcome and an overpressure applied to the salt solution to carry out the 'filtration'. Hence the term *reverse osmosis*.

The rate of permeation, ϕ, for reverse osmosis is given by:

$$\phi = A(\Delta P - \Delta \Pi)$$

where A is the permeation constant, ΔP is applied pressure, and $\Delta \Pi$ is osmotic pressure difference $(\Pi_W - \Pi_P)$.

The osmotic pressure of a 3 per cent salt solution is approximately 20 atmospheres, and reverse osmosis processes operate at pressures typically of 28 atmospheres.

Because the concentration of solutes builds up at the wall as in ultrafiltration (polarization), an added resistance to permeation occurs which must be included in the permeation expression. This resistance is normally included as part of the transfer constant A.

Conditions in the region of a membrane with a polarization (gel) layer are shown in Fig. 7.6.

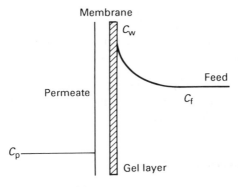

Fig. 7.6 Membrane with a gel layer.

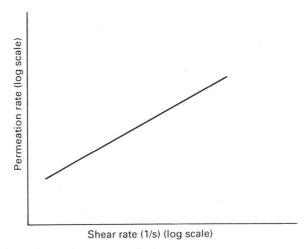

Fig. 7.7 Effect of shear on permeation rate.

In order to reduce the effect of the gel layer, a high shear rate at the surface of the membrane is used (generated by high velocities of fluid) to reduce the thickness of the gel layer, hence reducing the resistance to permeation. This effect is shown in Fig. 7.7.

The most satisfactory systems are those with a small cross-section flow channel giving high turbulence and hence high shear.

Freeze Concentration

If a solution of a liquid in water is carefully cooled to just below 0 °C, crystals of ice will start to form. These ice crystals will be surrounded by a solution of both liquids, but this solution will have been concentrated. If the ice crystals are separated from the crystal magma, then a concentrated solution is obtained. One difficulty with this process is that the ice crystals cannot be washed, because the ice will tend to melt thus diluting the concentrated solution.

The most successful systems are either those which employ a very efficient heat exchanger (mechanical, scraped surface type) giving a high level of control over the formation of ice crystals, or operate on a circulating tower basis with the ice crystals forming a 'bed' in the base; withdrawal of 'slush' for separation is balanced by a feed of chilled solution.

Adsorption/Desorption Processes[6]

These techniques use some physical, reversible characteristic to either retain temporarily or retard the passage of solute species. Examples of such operations are:

- *adsorption*. This operation makes use of physical forces (for example van der Waal's forces) to bind solutes to a solid phase, but as a reversible process. Typical solids used for adsorption processes are alumina, silica gel, activated carbon, etc.
- *ion exchange*. This process uses the phenomenon of different electrolytic forces between the solute molecules (electrolytes) and the ion exchange radicals on the resin.
- *gel filtration*. This process separates molecules on the basis of physical size by making use of the ability of molecules to penetrate the pores in the solid packing material.

All of the above techniques use packed columns in operation, and work on an essentially intermittent (batch) basis.

In the operation of a single column, on startup, the initial layers of the solid adsorb the solute until they become saturated and can hold no more solute. This saturation of the solid gradually extends to the whole of the solid bed in the column, and when this occurs, the liquid flowing out of the column will have the same composition as the feed.

The feed is then stopped and changed to the eluting liquid which washes the adsorbed solute from the solid. If the flowrate of eluting liquid is low, the concentration of liquid leaving the column will be high. Once all the solute has been removed from the solid, the column is ready to accept the feed solution again. With some solid adsorbents, it may be necessary to regenerate the solid before re-admitting the feed.

In order to obtain a continuous flow of separated solute, it is necessary to operate more than one column, and most systems use the columns in the following sequence:

Column 1:	(i) adsorbing	(ii) eluting	(iii) regenerating
Column 2:	(i) eluting	(ii) regenerating	(iii) adsorbing
Column 3:	(i) regenerating	(ii) adsorbing	(iii) eluting

These sequences are successively followed for each column.

Extraction

Extraction involves the use of a liquid to separate either:[6]

- a valuable solute from a solid material, or
- a solution of two liquids.

SEPARATION OF SOLID MIXTURES

Examples of this type of solid/liquid extraction process (alternatively known as leaching) are the extraction of vegetable oils from oil-bearing seeds using hexane (rape, maize, etc.) and also the extraction of mineral salts from ores. This particular operation is used mainly in mineral processing and in the food industry, and is rarely used in biotechnology.

SEPARATION OF LIQUID SOLUTIONS

This process depends on the formation of two phases when the extracting solvent is mixed with the original solution, and depends on choosing an extracting solvent which is essentially insoluble with the original solvent but will dissolve the solute. Acetic acid–water solutions, for example, can be separated using di(1-methylethyl)ether (i-propyl ether), the ether being insoluble in water but acting as a solvent for the acetic acid.

Penicillin can be extracted from water solutions using either amyl or butyl acetate, the basic process being shown diagrammatically in Fig. 7.8.

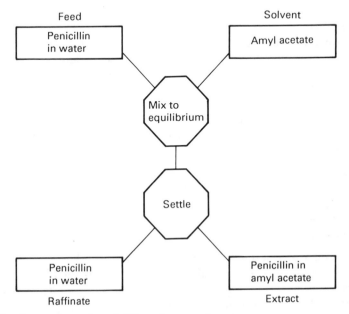

Fig. 7.8   Extraction of penicillin using amyl acetate.

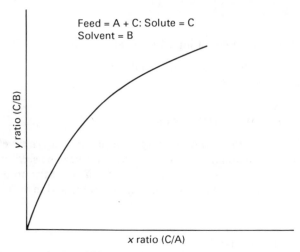

Fig. 7.9   Rectangular equilibrium distribution diagram.

The ratio of penicillin/water in the raffinate is different from this ratio in both the feed and extract, thus separation has taken place. The raffinate could be contacted with more amyl acetate, and further separation would then take place. If successive raffinates were treated over many stages, eventually almost all of the penicillin could be extracted into the amyl acetate, giving virtually complete separation of the original penicillin solution.

Equilibrium Data
The information required to design an extraction process is data for the equilibrium distribution of the solute between the original and extracting solvents.

The difficulty is that we have to deal with three components, and analytical (mathematical) representation would involve the use and solution of a number of differential simultaneous equations, which, although possible, is not simple.

It is much more convenient in these cases to use graphical methods of representation of data, and two basic systems are used:

- *triangular diagrams*, which are useful if the two solvents exhibit some appreciable solubility in each other
- *rectangular diagrams* plotting the ratio of solute/extracting solvent against the ratio of solute/original solvent (which can be used even if the two solvents are slightly soluble in each other).

A rectangular plot of solute/solvent ratios is shown in Fig. 7.9.

If the extracting solvent is carefully chosen so that it is essentially insoluble in the original solvent, *all* of the extracting solvent will appear in the extract and *all* the original solvent will appear in the raffinate.

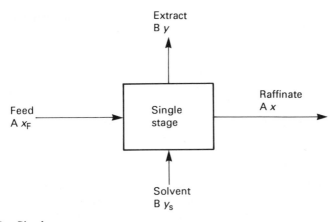

Fig. 7.10 Single stage contact.

Three types of process operation are commonly used:

- single-stage contact
- multiple-stage, cross contact
- multiple-stage, continuous, counter-current contact.

BATCH SINGLE-STAGE CONTACT

The process is represented schematically in Fig. 7.10, and represents the extraction of a solution containing original solvent (A) and solute (C) by extracting solvent (B).

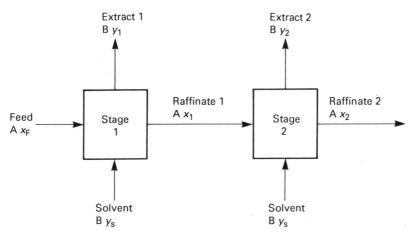

Fig. 7.11 Multiple-stage cross contact.

Assuming that A and B are essentially insoluble, then *all* of A will appear in the raffinate and *all* of B will appear in the extract. Let

x_F = ratio (C/A) in the feed;
y_S = ratio (C/B) in the solvent;
x = ratio (C/A) in the raffinate;
y = ratio (C/B) in the extract.

Then if A is the quantity of A *only* in the feed and B is the quantity of B *only* in the extracting solvent, an overall materials balance for the solute (C) will be:

$$Ax_F + By_S = Ax + By$$

This is the equation of a straight line, slope A/B passing through the points (x_F, y_S) and (x, y). This line (an *operating* line) can be constructed on the equilibrium diagram depending on the known characteristics of the system.

- if A, B, x_F and y_S are known, x and y can be found
- if A, x_F and y_S are known, and x and y specified, the quantity of solvent B can be obtained.

This construction is shown in Fig. 7.12.

MULTIPLE-STAGE CROSS CONTACT

A schematic diagram of this type of operation is shown in Fig. 7.11. It is more efficient to contact the original solution with multiple smaller quantities of solvent than to perform the operation in one stage, since more of the solute can be separated.

For this type of contact, the solute balances are as follows:

stage 1: $Ax_F + By_S = Ax_1 + By_1$
stage 2: $Ax_1 + By_S = Ax_2 + By_2$

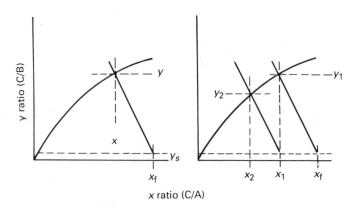

(a) Single stage   (b) Multiple stage cross-contact

Fig. 7.12   Construction for batch operation.

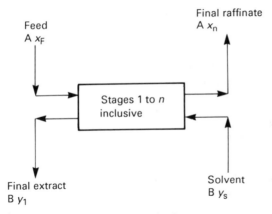

Fig. 7.13 Continuous counter-current contact.

Both of these equations are similar to the operating line for a single stage process, and the construction is the same for both stages (see Fig. 7.12). There is no reason why the quantity of solvent should not be different for both stages, and these quantities can be chosen to give a desired end result.

CONTINUOUS COUNTER-CURRENT CONTACT

This type of process is shown in Fig. 7.13, and a solute balance over *all* of the stages (1 to n inclusive) is given by:

$$Ax_F + By_S = Ax_n + By_1$$

This is the equation of a straight line of slope A/B but passing through (x_F, y_1) and

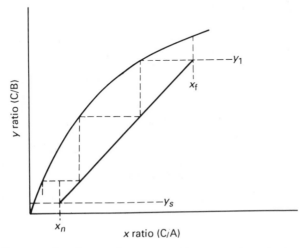

Fig. 7.14 Construction for continuous counter-current contact.

Table 7.4   Summary of notation used in Chapter 7

A	cross-sectional area	m^2
	quantity or flowrate of original solvent	kg; kg/s
B	quantity or flowrate of extracting solvent	kg; kg/s
C	solute concentration	kg/kg
$\mathscr{D}$	diffusivity coefficient	m^2/s
ΔP	pressure difference across membrane	N/m^2
R	retention/rejection factor	—
R_M	flow resistance of membrane	m^{-1}
R_G	gel layer resistance	m^{-1}
x	ratio of solute/original solvent	
y	ratio of solute/extracting solvent	
ϕ	rate of permeation through membrane	$m^3/s\,m^2$
μ	fluid viscosity	kg/m s
$\Delta\Pi$	osmotic pressure difference	N/m^2

(x_m, y_s). The construction for this operation is shown in Fig. 7.14, the number of stages required for the operation are stepped off on the diagram as shown. It should be noted that the term 'stages' used above refers to 'theoretical' or 'ideal' stages of 100 per cent efficiency in reaching equilibrium. Account must be taken of the fact that the efficiency of real stages will be less than 100 per cent.

A summary of the notation used in this chapter is shown in Table 7.4.

References

1. W.L. McCabe and J.C. Smith, *Unit Operations of Chemical Engineering*, 3rd edn, McGraw-Hill Kogakusha, Tokyo (1976).
2. J.E. Bailey and D.F. Ollis, *Biochemical Engineering Fundamentals*, 2nd edn, McGraw-Hill, New York (1986).
3. F.A. Glover, *Ultrafiltration and Reverse Osmosis for the Dairy Industry*, Tech. Bull. 5, National Inst. for Research in Dairying, (NIRD), Reading (1986).
4. P. Meares (ed.), *Membrane Separation Processes*, Elsevier, Amsterdam (1976).
5. (a) G. Hagen, 'Uber die Bewegung des Wassers in engen cylindrischen Rohren', *Ann. Phys.* (*Pogg. Ann.*) **46**, 423 (1839).
   (b) J. Poiseuille, 'Recherches experimentales sur le mouvement des liquides dans les tubes de tres petit diamètre', *Inst. de France Acad. des Sci. Memoires divers savantes* **9**, 433 (1846).
6. R.E. Treybal, *Mass Transfer Operations*, 3rd edn, McGraw-Hill, New York (1985).
7. Dominick Hunter Filters Ltd, Birtley, Co. Durham. Publications references (a) 502/2/84/UK, (b) 504/4/85/UK, (c) 516/12/86/UK.

Chapter 8

Manufacture of Solid Products

Drying

The drying of solids can be carried out in three ways:

- *Air drying*. The use of heated air (or inert gas) to provide the heat required for the vaporization of water. The air also acts as a 'carrier' for the removal of vapour from the system.
- *Contact drying*. The use of a heated surface in direct contact with the wet solid. Vaporization takes place within the solid, the vapour diffusing through the solid to the unheated surface. Unheated air at ambient conditions is frequently used to carry away the water vapour.
- *Vacuum drying*. The use of low pressures to vaporize the water at low temperatures and removal of the vapour via the vacuum system.

AIR DRYING[1, 2]

Basic Concepts
A typical plot of the rate of drying of a solid against the moisture content is shown in Fig. 8.1.

Moisture Content
There are two common methods of expressing the moisture content of a wet solid:

- '*Wet*' basis, expressed as a fraction of (water content) / (water content + solid). This is the most common analytical expression of concentration.
- '*Dry*' basis expressed as a ratio of water content/dry solid.

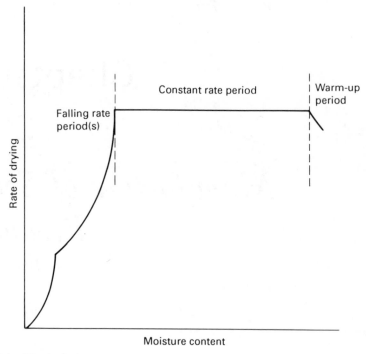

Fig. 8.1   Typical air drying characteristics.

In drying, the most common method used to express moisture content is the 'dry' basis, but the characteristic curve shown in Fig. 8.1 is of the same form irrespective of the basis used for moisture content.

Constant Rate Period
The rate of drying in this period is equivalent to the rate of evaporation from a pool of water. The surface of the solid is completely wet, and the rate of diffusion of water from inside the solid to the surface is the same as the rate of evaporation.

Falling Rate Period(s)
As the solid surface dries out and becomes 'patchy', moisture must be transported from the inside of the solid to the surface before evaporation can take place. The rate-controlling mechanism then becomes this transport of moisture from the inside of the solid to the surface, and the overall rate of drying falls.

Equilibrium Moisture Content
Although most solids can be dried to low moisture contents, some moisture will remain within the solid held by molecular forces. This is sometimes referred to as

bound moisture, and represents the equilibrium moisture. The equilibrium moisture content is a function of the temperature and humidity of the surrounding air.

THE AIR/WATER SYSTEM

The provision of heat and the removal of water using heated air depends on the physical characteristics of the air/water system. The important factors are the temperature variation of:

- humidity (ratio of water to dry air)
- percentage humidity (expressed as a fraction of saturation humidity)

These properties are most conveniently expressed in graphical form on a *psychrometric chart* similar to that shown in Fig. 8.2.

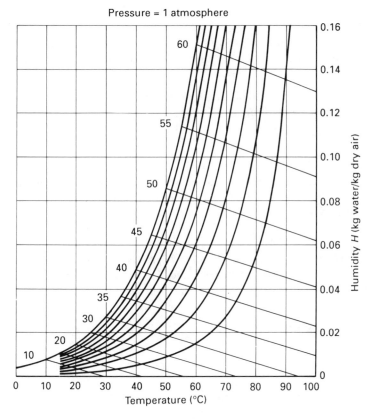

Fig. 8.2 Psychrometric chart for the air/water system.

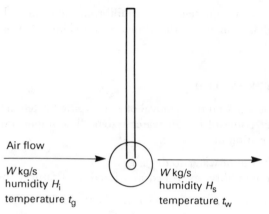

Fig. 8.3 Thermometer with wetted wick surround.

Wet Bulb Temperature

If the bulb of a thermometer is surrounded by a wet fabric (as shown in Fig. 8.3), then a heat and materials balance can be performed once a steady state has been reached.

- *Heat balance.* The air will pick up some water due to evaporation, say W kg/s, and the heat loss will be

$$q = W\lambda \tag{8.1}$$

$$= hA(t_G - t_W) \tag{8.2}$$

  where λ is the latent heat of vaporization, h is the heat transfer coefficient, and A is the area available for heat transfer (surface area).
- *Materials balance*

$$W = k_G A(H_W - H_G) \tag{8.3}$$

  where H_W is the humidity of the air at t_W, H_G is the humidity of the air at t_G, k_G is the mass transfer coefficient, and A is the surface area.

  Combining Equations (8.1) to (8.3) will give

$$k_G(H_W - H_G) = \left(\frac{h}{\lambda}\right)(t_G - t_W)$$

or

$$(H_W - H_G) = \left(\frac{h}{k_G\lambda}\right)(t_G - t_W)$$

For the air/water system, provided that the air velocity flowing over the wetted wick exceeds 5 m/s, the term h/k_G is constant (s), and

$$(H_W - H_G) = \left(\frac{s}{\lambda}\right)(t_G - t_W)$$

or

$$(H - H_S) = -\left(\frac{s}{\lambda}\right)(t_D - t_W) \tag{8.4}$$

where H is the humidity of the air at temperature t_D, H_S is the humidity of the air at saturation, t_D is the dry bulb temperature, and t_W is the wet bulb temperature.

Equation (8.4) is the equation of a straight line of slope $-s/\lambda$ relating H to t_D, t_W, and H_S, and is known as an *adiabatic cooling line*.

On the psychrometric chart, Equation (8.4) represents a family of lines which are theoretically parallel, but since the latent heat of vaporization λ is a function of temperature, the scales of the chart are slightly distorted to make the adiabatic cooling lines parallel for ease of interpolation. The adiabatic cooling lines have been included in Fig. 8.2.

During an air drying process, air is contacted with the wet solid and saturates along the appropriate adiabatic cooling line. A typical process uses air at ambient conditions, heats it (with no change in moisture content, i.e. humidity) and on bringing it into contact with the solid, it saturates to H_S along the appropriate adiabatic cooling line. The process is represented diagrammatically on the psychrometric chart in Fig. 8.4.

The value of $(H_S - H)$ represents the capacity of the air for water removal in kg water/kg dry air, and it can be seen from Fig. 8.4 that the same quantity of air (initially at t_1) heated to a higher temperature than t_2 will have a better capacity for drying than air at t_2.

Temperature During Drying
During the constant rate period, the solid attains a temperature which approximates to the wet bulb temperature (t_W). During the falling rate period, the temperature of the solid steadily rises until the dry bulb temperature (t_D) is reached when the solid is completely dry.

Drying is frequently operated as a multi-stage rather than a single-stage process in order to maintain the temperature of the air in direct contact with the solid at a reasonable level. This is desirable particularly when drying heat-sensitive biological materials.

Comparison of a multi-stage process with a single-stage process is shown in Fig. 8.5. Both processes are designed to achieve the same drying result, but it can be seen that in the three-stage process that the maximum temperature of the air (and the product) would be 60 °C. In a single stage of contact the temperature would be considerably in excess of this (approximately 123 °C), with a stronger possibility of damaging a thermally sensitive product.

Non-porous materials (gels, etc.) can be damaged during the falling rate period when cracking of the surface may occur, and the rate of drying must be carefully

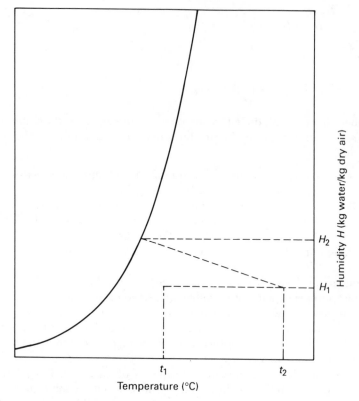

Fig. 8.4 Capacity of air for drying.

controlled. In cases of this type, co-current air flow can have some advantages, even though from a heat and mass transfer point of view the process may not be optimal.

Drying of Suspensions or Solutions
For these materials, conventional belt or oven dryers are unsuitable, and the best type of dryer to use is the *spray dryer*. In a spray dryer, the suspension or solution is atomized into small droplets and then contacted with the heated air. A schematic diagram of this type of system is shown in Fig. 8.6.

It should be noted that the air flow in Fig. 8.6 is co-current to the flow of atomized liquid. In dryers of this type it is not unusual for the inlet air to be at a temperature of 200 °C or more, and co-current operation ensures that the hottest and driest air comes into contact with the wettest material (the temperature of the material approximating to the wet-bulb temperature of the air). In this way the material receives minimal thermal damage because of the short residence time in the dryer.

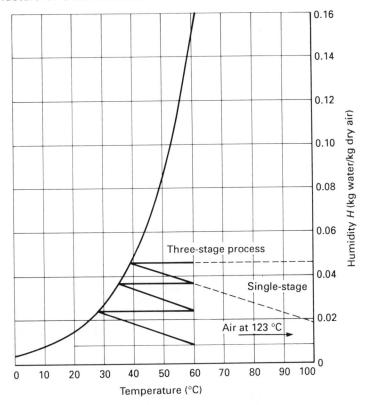

Fig. 8.5 Comparison of single and multi-stage drying processes.

Atomization can be achieved in a number of ways[1] using:

- pressure jets
- spinning cups
- spinning discs.

CONTACT DRYING[2]

The mechanism of heat transfer in contact drying is due to conduction of heat through the wet material which is in contact with a heated surface (Fig. 8.7). Unsteady-state heat and mass transfer take place during the drying process and this means that the solution of the mass transfer problem is mathematically more complex than in air drying.

The average moisture content of the material during contact drying, assuming an infinite slab of material is given by[3]

$$\frac{(W - W_E)}{(W_0 - W_E)} = \left(\frac{8}{\pi^2}\right) \exp\left[-D\theta\left(\frac{\pi}{2d}\right)^2\right]$$

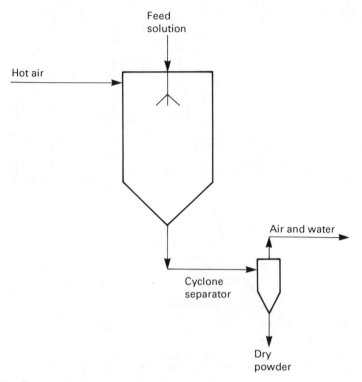

Fig. 8.6 Schematic diagram of a spray drying system.

where W is the moisture content at time θ, W_E is the equilibrium moisture content, W_0 is the initial moisture content, D is the diffusivity coefficient, and d is the half-thickness of the slab.

This method of drying is suitable for comminuted material (e.g. fishmeal) where the rate of drying is quite high. It is also used for roller drying of materials like milk, cereals, baby foods, etc., using either a roller dryer as shown in Fig. 8.8, or a rotary dryer containing heated tubes.

VACUUM DRYING[2]

By reducing the pressure of the system during a drying process, the temperature of the material at the end of the drying process can be reduced due to the decreased boiling point of water at low pressures (see p. 90).

One of the disadvantages of the removal of water from a biological material is that as the water is removed, concentration of salts takes place and causes water migration *within* the cells. This migration of moisture due to osmotic forces as a result of the increase in salt concentration, can cause cell-wall disruption.

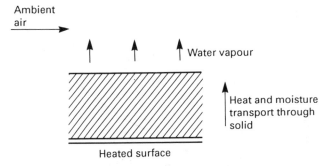

Ambient air

Water vapour

Heat and moisture transport through solid

Heated surface

Fig. 8.7 Contact drying of wet material.

Freeze Drying[4-6]

The phase diagram for water is shown in Fig. 8.9. For water at point A on the diagram, a reduction in pressure to point B will vaporize the water without an increase in temperature. However, there will be a potential migration in moisture due to salt concentration in the cells setting up osmotic forces.

By freezing the solid (from point A to point C on Fig. 8.9) and then reducing the pressure (to point D), the ice will *sublime* directly into vapour without passing through the liquid phase. Since the water remains as a solid, salt migration cannot take place, and thermal and osmotic damage are eliminated.

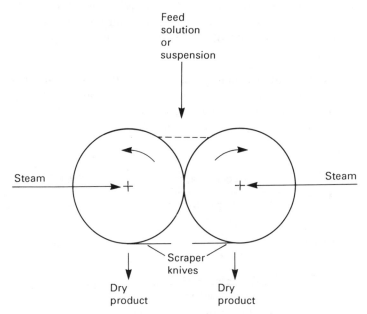

Feed solution or suspension

Steam   Steam

Scraper knives

Dry product   Dry product

Fig. 8.8 Roller drying system.

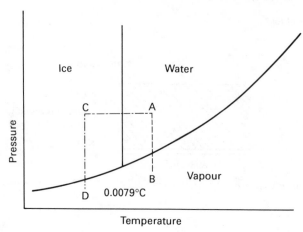

Fig. 8.9   Phase diagram for water.

The rates of sublimation *at atmospheric pressure* are low due to the low vapour pressure of ice (*ca.* 1 mmHg), but around the vapour pressure of ice, the rate of sublimation is high, and optimum operation takes place at about 0.5 to 1.0 mmHg absolute pressure.

For a freeze drying process we must provide the following:

- an evacuated chamber (0.5–1.0 mmHg abs)
- some means of providing the *latent heat of sublimation* $(2.8 \times 10^7 \, \text{J/kg})$
- provision for the adequate removal of large volumes of water vapour and its condensation (at 1.0 mmHg the specific volume of water vapour is $1100 \, \text{m}^3/\text{kg}$).

Heat is usually provided by contact surfaces which allow the sublimed vapour to escape, and no melting should occur during the process. Expanded metal (steel, aluminium) is usually used, as shown in Fig. 8.10, the heat being conducted through the expanded metal into the frozen solid.

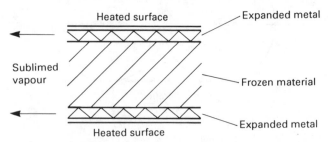

Fig. 8.10   Expanded metal used in freeze drying.

Table 8.1 Salting-out order of ionic groups (in decreasing order)

Anions		*Cations*	
citrate	$(3-)$	aluminium	$(3+)$
tartrate	$(2-)$	magnesium	$(2+)$
sulphate	$(2-)$	ammonium	$(+)$
acetate	$(-)$	potassium	$(+)$
chloride	$(-)$	sodium	$(+)$

Precipitation

The precipitation of proteins is achieved by changing their solubility characteristics, usually in two ways:

- by using ionic inorganic salts
- by the addition of organic solvents.

It is also possible to remove water from solutions by the addition of non-ionic polymers and flocculating agents, thus forcing precipitation by effectively removing water molecules from solution by adsorption onto the polymer and 'concentrating' the solution.[7]

SALTING OUT

Table 8.1 shows the salting-out effect of various ions (both cationic and anionic) and is arranged in diminishing order of salting-out effect, thus citrate ions have a large effect, chloride ions the least effect. This table is not exhaustive, and there are other ions which exhibit precipitation effects.

The effect on the solubility of proteins by an ionic group is dependent on the ionic strength of the solution. The solubility (S) versus ionic strength (P) usually follows the relationship:

$$\log S = B - K_S P$$

S is solubility (g/L), B is a constant, P is ionic strength, and K_S is the salting-out constant.

The ionic strength P is given by

$$P = 0.5\Sigma(C_i Z^2)$$

where C_i is the molar concentration of ionic species, and Z_i is ionic valency.

Example 8.1
Calculate the ionic strength of a 1 M sodium sulphate solution.

Answer
1 mol of Na_2SO_4 contains

(a) 2 mol ions of sodium (Na) at ionic valency 1
(b) 1 mol ion of sulphate (SO_4) at ionic valency 2.

Thus

$$P = 0.5[(2 \times 1^2) + (1 \times 2^2)] = 3$$

Figure 8.11 shows a typical plot of solubility against ionic strength for a number of ions. The use of this method, for example the addition of sodium sulphate to an enzyme solution, will precipitate the enzyme and give a solid with very strong enzyme activity. Unfortunately this solid is deposited as a 'sludge' and separation by filtration is almost impossible, and may even present difficulties using a centrifuge.

Because of this separation difficulty, it is common practice to use another solid which will filter easily to 'carry' the precipitated protein solids. Thus, for example, the addition of diatomaceous earths or cellulose before precipitation makes the subsequent solid/liquid separation easier. This method of production is successful provided that the added solid is acceptable in the final product.

Another potential problem is that the precipitated solid will also contain some of the ionic salt used to promote precipitation. It is impossible to wash out this salt, otherwise resolubilization of the protein would occur.

ORGANIC SOLVENTS

The addition of alcohols and ketones to protein solutions causes the protein to precipitate, mainly due to changes in the dielectric properties of the solution. If the process is carried out at *ca.* $-5\,°C$, only small amounts of solvent are required.

Common industrial practice is to use one of the higher alcohols, such as i-propanol (propan-2-ol) which is readily available at low cost. Operation at

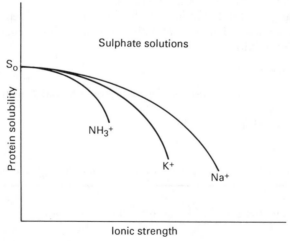

Fig. 8.11   Solubility *vs* ionic strength.

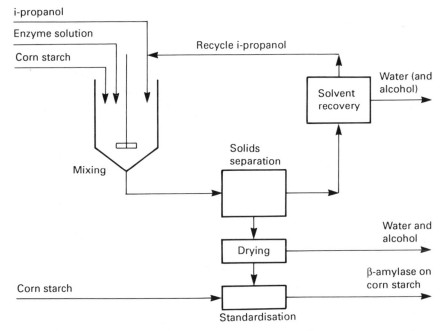

Fig. 8.12 Precipitation of β-amylase on to corn starch.

−5 °C requires refrigeration, which is expensive, and the i-propanol is usually added at about room temperature.

The overall economics of the process indicate that a temperature of 10–15 °C is optimum. This is true provided recovery and recycle of the i-propanol is included as part of the process, otherwise solvent costs could cause the process to become uneconomic.

The protein is usually precipitated in the presence of a carrier solid in order to ease subsequent separation. The alcohol remaining in the solid after separation is readily removed by a simple drying process.

Figure 8.12 is a schematic diagram of the process used for the precipitation of β-amylase using corn starch as a carrier.

Crystallization[1, 8]

Crystallization of a solute from a solution is used for the purification of the solute as well as providing a solid crystalline product.

CRYSTALLIZATION OF SALTS

The solubility of salts in water is a function of temperature (see Fig. 8.13), and the purification of materials like common salt (NaCl), sugar, and other inorganic salts

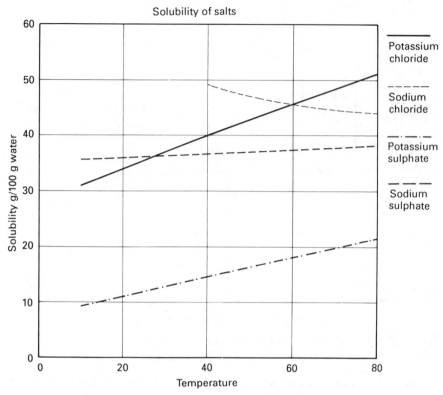

Fig. 8.13   Solubility of salts with temperature.

requires the production of a saturated solution at an elevated temperature, followed by cooling to force precipitation (crystallization) of the salt from the concentrated solution.

It will be seen from Fig. 8.13 that sodium sulphate exhibits a reverse solubility curve (solubility decreases with temperature).

The crystals can be separated from the solution using filtration, etc., and the separated *mother liquor* recycled back to the concentration stage. A schematic diagram of the basic process is shown in Fig. 8.14.

The rate of crystallization is dependent on the degree of supersaturation produced on cooling, and the crystal size is a function of the rate of crystallization (the more rapid the rate, the smaller the crystals). The concentration of impurities in the solution also affects the size of the crystals, and in order to prevent a build-up of impurities due to recycling the mother liquor, it is normal practice to reject a portion of the recycle. This is the 'bleed' stream shown in Fig. 8.14.

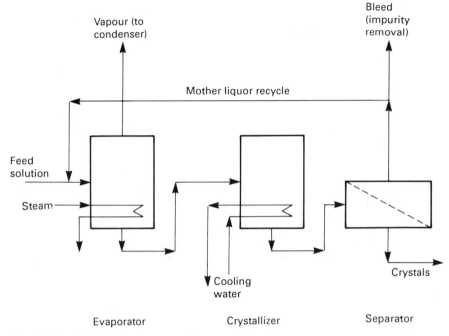

Fig. 8.14 Schematic diagram of a crystallization process.

ANTIBIOTICS

The crystallization of antibiotics could be carried out in a similar manner to that of inorganic salts. However, a concentration stage is required to produce the saturated solution (usually using vacuum evaporation), and with some antibiotics temperatures less than 15 °C can destroy the antibiotic activity.

In the case of a heat-sensitive antibiotic like penicillin G, crystallization (and purification) takes place by the addition of ionic salts and alcohol (combined *salting-out* and dielectric interference). The penicillin is extracted into amyl (or butyl) acetate which has the effect of some purification, as well as concentration, and this stage is followed by the addition of sodium acetate and ethyl alcohol. Crystals of penicillin G (sodium salt) are produced which are then filtered or centrifuged out from the solution. The total operation takes place at about 0 °C, thus ensuring minimal thermal damage.

A summary of the notation used in this chapter is shown in Table 8.2.

Table 8.2 Summary of notation used in Chapter 8

A	area available for heat (or mass) transfer	m^2
C_i	molar concentration of ionic species i	$mol\ ions/m^3$
H	humidity of air	kg/kg dry air
H_S	saturation humidity	kg/kg dry air
h	heat transfer coefficient	$W/m^2\ K$
K_S	salting-out constant	$m^3/kmol$
k_G	mass transfer coefficient	$kmol/s\ m^2$ atmos
P	ionic strength	$kmol/m^3$
q	rate of heat transfer	$W\ (J/s)$
S	solubility of salt	kg/m^3
t	temperature	C
t_D	dry bulb temperature	C
t_W	wet bulb temperature	C
W	rate of evaporation	kg/s
Z_i	ionic valency	—
λ	latent heat of vaporization	J/kg

References

1. J.M. Coulson, J.F. Richardson, J.R. Backhurst, and J.H. Harker, *Chemical Engineering*, Vol. 2, 3rd edn, Pergamon Press, Oxford (1978).
2. R.B. Keey, *Drying Principles and Practice*, Pergamon Press, Oxford (1972).
3. A.B. Neuman, 'The drying of porous solids: diffusion calculations', *Trans. AIChemE*, **27**, 310 (1951).
4. H.G. Maister, E.N. Heger, and W.M. Bogard, 'Continuous freeze drying of *Sematia marcescens*', *Ind. Eng. Chem.* **50**, 623 (1958).
5. J.C. Harper, C.O. Chichester, and T.E. Roberts, 'Freeze drying of foods', *Agric. Eng.* **43**, 78 (1962).
6. I.D. Mellor, *Fundamentals of Freeze Drying*, Academic Press, London (1978).
7. J.E. Bailey and D.F. Ollis, *Biochemical Engineering Fundamentals*, 2nd edn, McGraw-Hill, New York (1986).
8. J.W. Mullin, *Crystallisation*, 2nd edn, Butterworths, London (1972).

Chapter 9

Large-scale Industrial Processes

This chapter has been written with the object of describing a number of large-scale biotechnology processes, in order to illustrate the material contained in previous chapters. It is not intended to be a comprehensive treatise covering all processes used industrially, neither is it intended to cover every small detail of the processes selected.

The processes described here have been chosen to cover the range of industrial operations in common use. For example, aerobic and anaerobic fermentation; severe upstream processing requirements (mainly in terms of sterility); and severe and complicated downstream processing requirements.

In each case, the operations involved have been separated into convenient areas: upstream and downstream processing and the fermentation stage.

Penicillin Manufacture[1, 2]

The development of penicillin as an antibiotic agent started in the early 1940s, the incentive for development being World War II. Initial production was by static culture using glass milk-bottles, but the submerged (deep) fermentation process had been developed by 1945, and world-wide production has been carried out using this type of process since that time.

Penicillin was the first antibiotic to achieve large-scale production, and much development on both the biology and downstream processing has been carried out over the last 40 years. Many of the developments in fermentor design, carried out to achieve an optimum penicillin production rate, have been of immense value in areas of biotechnology other than antibiotic manufacture.

The original strain of mould used was *Penicillium notatum*, which was quickly superseded by the *P. chrysogenum* strains now used exclusively in the production of penicillin and its derivatives.

SUBSTRATE AND FERMENTATION KINETICS

The biology of the penicillin production process has been extensively researched, and the modern process is a two-stage fermentation in which the *P. chrysogenum* is initially cultivated under conditions designed to achieve rapid mycelial growth, followed by the production of penicillin during a slow mycelial growth phase.

This reaction pattern has led to the use of a 'fed-batch' process, where the fermentor is operated basically in batch mode, but after growth has reached a certain point, a continuous (small) feed of extra nutrients is started, and continued until the penicillin titre reaches the required level. Once this point is reached, the reaction is terminated and the 'batch' of material sent for processing and recovery.

Batch Fermentation

The main nutrients for this process are glucose, lactose and corn steep liquor (obtained as a waste product from the manufacture of corn starch). During the initial stages glucose/corn steep carbon is consumed giving rapid growth and high mycelium content. Following this rapid growth phase, lactose utilization takes over, giving slow mycelial growth but producing high titres of penicillin. Penicillin G production is improved if the side-chain precursor phenyl acetic acid (or a derivative such as phenylalanine) is added at a concentration of $0.2–0.8 \, kg/m^3$ at the start of the fermentation. The oxygen transfer rate is important, an uptake rate of $0.3–0.5 \, (v/v)/min$ being required at a critical dissolved oxygen concentration of $0.022 \, mol/m^3$. The usual aeration rate is from 0.5 to 1.0 $(v/v)/min$ to ensure that the carbon dioxide produced by the organism is quickly flushed from the system. The pH of the substrate is controlled between 6.8 and 7.4.

Fed-batch Fermentation

Hexose monomers (like glucose) and derivatives (like sucrose) are added continuously in controlled amounts to optimize penicillin and biomass production. The control of the addition is important to prevent an accumulation of hexose (which inhibits penicillin production). Lactose (which is expensive) has now been completely replaced by (cheaper) glucose.

The continuous feed is started after initial glucose/corn steep carbon has been utilized during the rapid growth phase. The rate of addition of glucose is determined by the need to prevent autolysis, and by the oxygen demand rate.

Penicillin production is dependent on the pH of the substrate, with an optimum around 6.5 depending on the strain of *P. chrysogenum* being used.

The basic fermentor is still the agitated, sparged fermentor. The volume of the fermentor in the 1950s was $20 \, m^3$, but modern penicillin fermentors are of the order of $150–200 \, m^3$.

Table 9.1 Fed-batch operating variables

Volume	150–200 m^3
Power input	3–4 kW/m^3
Air flowrate	0.5–1.0 (v/v) min
Pressure	0.3–0.7 atmospheres overpressure
Temperature	25 °C (depending on strain)
pH	6.5 (depending on strain)
Fermentation time	180–220 h
Carbon utilization	6–10 per cent for penicillin production

To allow for the continuous feed of nutrients during the fed-batch operation, the initial filling of the fermentor at the start of operations does not exceed 80 per cent.

Operating Variables
The operating variables for the fed-batch process are given in Table 9.1.

Agitator speed is also an important operating variable: high speeds tend to produce dense pellets rather than filamentous mycelium, which reduces the apparent viscosity of the broth and consequently reduces agitator power requirements. However, diffusion of nutrients into the pellets, as well as diffusion of products out of the pellets, can become limiting, and agitator speeds are usually limited to form short mycelium filaments.

Inoculum Development
Calcium carbonate is used in the seed vessel to prevent low pH values, since the quality and quantity of inoculum affects the subsequent production of penicillin. Sudden breakdown of production in the main fermentor can be due to ineffective inoculum, and above 5×10^9 spores/m^3, filamentous mycelium with high enzyme contents and good penicillin production rates are observed. Below this spore level dense pellets tend to form and glucose utilization is less efficient, leading to poor penicillin production.

DOWNSTREAM PROCESSING

The basic processes involved in the recovery of penicillin from the fermentor broth have changed little since the early 1950s; only the scale has changed.

Figure 1.3 shows the general block diagram of the penicillin process, and Fig. 9.1 shows the processes in diagrammatic form.

Broth Filtration
The separation of the mycelium from the fermentor broth is usually carried out continuously using a rotary vacuum filter, the mycelium being washed on the drum. The composition of the original fermentor medium has a marked influence on the filtering characteristics of the broth, protein constituents leading to

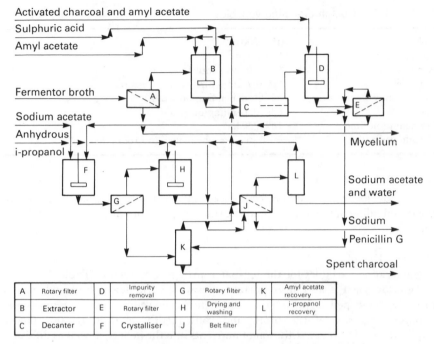

Activated charcoal and amyl acetate

Sulphuric acid

Amyl acetate

Fermentor broth

Sodium acetate

Anhydrous
i-propanol

Mycelium

Sodium acetate
and water

Sodium

Penicillin G

Spent charcoal

A	Rotary filter	D	Impurity removal	G	Rotary filter	K	Amyl acetate recovery
B	Extractor	E	Rotary filter	H	Drying and washing	L	i-propanol recovery
C	Decanter	F	Crystalliser	J	Belt filter		

Fig. 9.1   Penicillin manufacture—downstream processing operations.

difficulties of separation. Most fermentor medium is formulated not only to give good penicillin production rates but to give easy filtration characteristics.

As an alternative to filtration, it is possible to send the mycelium suspension directly to the extraction stage (whole broth extraction), but extra impurities can be leached from the mycelium giving problems in later processing. There is also a waste disposal problem with the separated mycelium. Washed, filtered mycelium (penicillin-free) has a potential market as a cattle feed or fertilizer.

Liquid Extraction
The penicillin in the filtrate from the broth separation is extracted in the acid form into amyl or butyl acetate. In order to reduce the damage to penicillin due to the acid conditions, the control of temperature, pH, and sterility is critical. The extraction is carried out at 0–3 °C and pH 2.5–3.0 (using sulphuric or phosphoric acid). Because the degradation of penicillin in acid conditions is a time/pH relationship, special types of extractors have been developed to give a short contact time at these low pH values. The extractor is a centrifugal device designed to give:

   • rapid intimate contact of the acid–penicillin with the acetate solvent, and
   • rapid separation of the two phases by centrifugal action.

Because the vigorous mixing of the two phases can lead to the formation of stable emulsions (which makes separation almost impossible), it is common practice to add anti-emulsion compounds in the acid–penicillin feed. The water phase containing small traces of penicillin and also some of the solvent, is sent to the solvent recovery stage.

Impurities Removal
Pigments and other trace impurities are removed from the acetate/penicillin phase by treatment with activated charcoal in a simple mixing vessel. After treatment, the charcoal suspension is separated, again using a rotary vacuum filter. During this filtration the cake is washed with more solvent. The charcoal cake, containing some solvent, is also sent to the solvent recovery plant.

Crystallization
The addition of sodium or potassium acetate to the solvent/penicillin phase causes the solubility of the penicillin to decrease, and the sodium or potassium salt crystallizes out. Careful control of the concentration of sodium/potassium with respect to the penicillin concentration, and of the pH and temperature give an optimum yield and size of crystals. The crystals are separated using a rotary vacuum filter, the separated liquor being sent for recovery of solvent.

Crystal Drying
The crystals of penicillin are mixed with a volatile solvent (anhydrous i-propanol, butanol, or ethanol) which has the double effect of removing impurities and sodium/potassium acetate, and also performing some pre-drying. The crystal suspension is filtered and washed with more anhydrous solvent and dried using warm air. The crystal product is about 99.5 per cent pure penicillin, and is used either as a feedstock for further processing to pharmaceutical grade penicillin G, or as an intermediate for the production of semi-synthetic penicillin derivatives.

The solvent used at the drying stage is recovered by distillation, and the anhydrous solvent recycled back to the production process.

SYNTHETIC PENICILLINS

The synthetic penicillins all contain the 6-aminopenicillinic acid (6-APA) nucleus which is coupled with different radical groups to give a large range of semi-synthetic penicillins.

The 6-APA can be manufactured by either a fermentation process omitting the phenyl acetic acid side-chain precursor, or by using an enzymatic conversion of penicillin G. Acylation of the 6-APA using acid chlorides gives the synthetic molecules. Some examples of the semi-synthetic penicillins are shown in Fig. 9.2.

Effluent Treatment[3–6]

Effluent treatment refers to the treatment of waste liquid streams from either industrial processes, or from domestic sources, and is alternatively referred to as 'waste water treatment'.

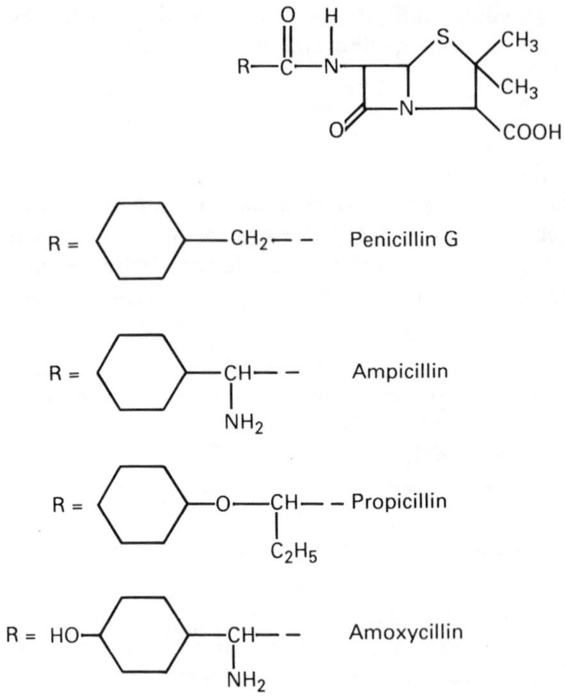

Fig. 9.2 Structures of some synthetic penicillins.

The aim of the treatment is to provide a liquid which is suitable for discharge into a river or a tidal estuary which is of such a composition that it has a minimal effect on the natural balance of the environment.

Contaminants from industrial sources can be a mixture of inorganic and organic chemicals (from chemical manufacturing or processes using chemicals, e.g. metal treatment), or natural substances like proteins, carbohydrates, etc. (from food processing or biotechnology operations). The treatment of any particular effluent is very specific depending on the range of materials contained in the effluent stream.

The general treatment scheme for an effluent (including domestic sewage) is made up of the following stages, the scheme of treatment being shown diagrammatically in Fig. 9.3:

(1) removal of large solids
(2) removal of fats and oils
(3) removal of suspended solids (fine particles)
(4) neutralization/removal of heavy metals
(5) removal of undesirable chemical ions
(6) adjustment of pH level
(7) reduction of the biological/chemical oxygen demand.

The major criteria used by the various national controlling bodies as a measure of the strength of an effluent are:

- *Biological oxygen demand (BOD)* (mg/L or ppm). The BOD requirement is the quantity of dissolved oxygen consumed by an effluent in a fixed time, normally a period of 5 days, when incubated at 20 °C
- *Chemical oxygen demand (COD)* (mg/L or ppm). The COD requirement is the quantity of dissolved oxygen consumed by an effluent in an acidic solution of potassium dichromate at *ca.* 80 °C, the analysis taking 2 h.

The COD analysis oxidizes more of the contaminating material than the BOD analysis, and the COD value is inevitably higher than the BOD value for the same sample. However, since the result of a COD analysis is known in 2 h compared

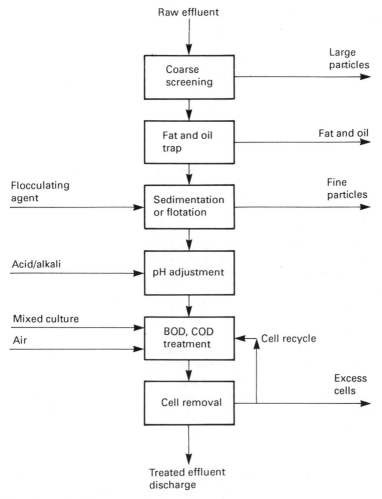

Fig. 9.3 General effluent treatment scheme.

Table 9.2 Typical strengths of raw effluents

Process operation	Main components	BOD value	COD value
Abattoir	suspended solids, protein	2600	4150
Domestic sewage	suspended solids, oil/grease, carbohydrates, protein	350	300
Beet sugar	suspended solids, carbohydrates	850	1150
Meat canning	suspended solids, fat, protein	8000	17 940
Dairy	carbohydrates fat, protein	600	—
Petroleum refining	phenols, hydrocarbons, sulphur compounds	850	1500
Fermentation		4560	4120
Starch reduction of flour	suspended solids, carbohydrates, protein	12 000	17 150
Brewing	carbohydrates, protein	10 000	16 500

with 5 days for a BOD analysis, the COD value is a better criterion to use for control purposes.

In recent years in industrialized countries, the restrictions on effluent discharge have steadily become more severe as efforts have been made to improve the natural environment. There is now national legislation covering a wide range of materials capable of causing pollution, and these must be reduced to acceptable levels before discharge takes place.

Table 9.2 gives some typical values of the parameters found in a number of raw (untreated) effluents, and Table 9.3 shows typical levels of contaminants which must be achieved before discharge takes place.

The BOD/COD level of a liquid effluent is increased by materials like starches, proteins, carbohydrates, etc. which find their way into the waste liquid streams during food processing operations, and also during biotechnology operations. The treatment of such wastes is usually carried out using a mixed culture of organisms chosen to break down the 'contaminants' and reduce the BOD/COD level.

PHYSICAL TREATMENTS

Large Solids
Large pieces of meat, skin, hides, feathers, etc. can be removed by using standard screening techniques, usually using a rotary perforated drum. This type of crude filter is suitable for the removal of solids having a size range down to 2–3 mm.

Table 9.3 Typical requirements for discharge

	River discharge	Sewer discharge
Suspended solids	30 ppm	1000 ppm
Temperature	30 °C	42 °C
pH	5–9	5.5–11 (10 min)
		6–10 (30 min)
Quantity		680 m³/day
		110 m³/h max
		225 m³/3 h max
Oil and fat	10 ppm	10 ppm
Chloride	1 ppm	1 ppm
Cyanide	0.1 ppm	0.1 ppm
Sulphate	1 ppm	1 ppm
Heavy metals	25 ppm	100 ppm
BOD	20 ppm max	68 g/day*

*Note: total BOD is often expressed as a quantity, e.g. 680 m³/day at 100 ppm BOD; weight of effluent = 680 × 1000 kg/day; BOD (at 100 ppm) = (680 × 1000) × 100 × 10⁻⁶ = 68 kg BOD/day.

Intermediate Solids
More sophisticated separation techniques are required if the solids are in the size range down to 100–120 μm, and pressure filters or rotary drum filters can be used.

Suspended Solids
Although centrifugation of suspended solids (size range 2–50 μm) is possible, the operation can be uneconomic due to the large volumes of effluent at low solids concentrations (1 per cent is typical). Such large volumes often require large centrifuges, and an alternative is to use either gravity settling devices or froth flotation techniques.

- *Gravity settling.* The device usually consists of a tank with sufficient cross-sectional area for flow to allow the small suspended particles time to settle to the bottom of the tank (velocities of 2 m/h). The settled solids can be continually removed from the base of the tank using paddle scrapers travelling at low velocity (less than 2 m/s) so as not to disturb the sediment. Alternatively the solids are allowed to accumulate over a period of weeks, and are then removed during plant shutdown. The system is similar to that shown in Fig. 9.4(a).
  Throughputs of the order of 50 m³/day per m² cross-sectional area with 2 h or more residence time allow up to 80 per cent solids removal from the effluent.
- *Froth flotation.* By introducing small bubbles of air into the effluent, the suspended solids become attached to the bubbles and the solid/air bubbles float to the surface. The solid/air froth is continually removed using scraper paddles. A typical layout is shown in Fig. 9.4(b).

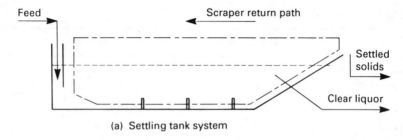

(a) Settling tank system

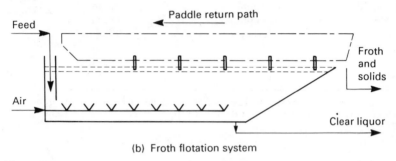

(b) Froth flotation system

Fig. 9.4 Suspended solids removal.

BIOLOGICAL TREATMENT

Oxygenation and Digestion

This technique makes use of the naturally occurring organisms in the effluent to aerobically decompose the biodegradable material present. A number of devices have been developed over the years, initially consisting of large shallow ponds filled with coarse material (coke, coal, crushed stone) where the effluent is pumped to the pond, allowed to digest for a period of time (2–12 h), then emptied. Whilst empty the bed is allowed to 'rest' for a further period of time (10 h typically) to allow the biologically active slimes coating the filling to recover and absorb oxygen from the atmosphere. The cycle begins again when the pond is refilled with effluent.

Because of the large land area required with what are essentially static ponds, the trickle bed filter was developed. Here the effluent is continuously fed to the top of the bed and air flows counter-currently through the bed by natural circulation due to the design of the bed shape. This speeds up the biodegradation and provides a better utilization of the space and packing. A trickle bed filter is shown schematically in Fig. 9.5.

Biological Digestion

This process differs from the natural digestion systems described above in that organisms are deliberately added to the effluent to reduce material with a high

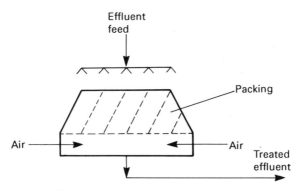

Fig. 9.5 Trickle bed filter.

BOD. A typical system is the 'activated sludge' process using a broad-band mixture of organisms selected specifically to treat the particular effluent.

The sludge and effluent are mixed and oxygenated in a treatment tank which is basically a continuous stirred tank fermentor (CSTF), sized to give the necessary residence time to reduce the BOD level to that required. The overflow from the digester is treated to separate the active sludge, which is recycled back to the CSTF. The sludge may be separated using either a continuous filter or centrifuge. A schematic diagram of an activated sludge process is shown in Fig. 9.6.

Anaerobic Digestion
Solid materials such as cellulose, farm waste, vegetable matter, and biomass can be treated anaerobically to produce a methane/carbon dioxide gas mixture containing about 70 per cent methane, which can be used as an energy source. This type of process is limited commercially to the digestion of the micro-organism sludge obtained from aerobic processing, in order to reduce the disposal problems.

Fuel Alcohol[7–11]

Ethanol has been produced biologically in the manufacture of alcoholic beverages for many centuries, and the use of ethanol in the internal combustion engine dates from the turn of the century.

One of the advantages of using fermentation for alcohol production is that the substrates for the fermentation can be readily obtained from renewable resources or waste material. The alcohol fermentation has been widely studied, and the fermentation and recovery technology already exists in a highly developed state.

Apart from the manufacture of alcoholic beverages, large-scale manufacture of ethanol by fermentation has concentrated on the use of alcohol as a replacement (or partial replacement) for petroleum-derived fuels, specifically for application in the internal combustion engine.

Ethanol, however, is also a useful starting material for the manufacture of many

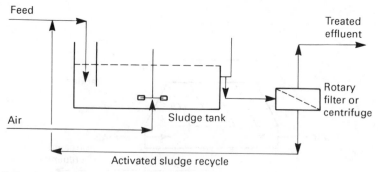

Fig. 9.6   Activated sludge process for effluent treatment.

chemicals, for example acetic acid, acetaldehyde, butanol, chloroform, and ethylene (which is itself the major intermediate in the petrochemical industry). Thus, ethanol produced from renewable sources by fermentation, together with methane and perhaps ethylene and hydrogen from biomass or wastes, could possibly be utilized to replace the petroleum feedstocks currently required by the chemical industry.

The largest programmes for the production of ethanol by fermentation are the Brazilian National Alcohol Programme started in 1975, and the US Gasohol programme, both conceived to reduce the national dependence on petroleum imports. Both programmes were initially targeted for the replacement of gasoline as a fuel for the internal combustion engine. Differences in philosophy exist between the two programmes; the Brazilian concept is to replace completely the use of gasoline by ethanol (and eventually to replace petroleum-derived diesel fuel) on national economic grounds; the aim in the USA is to use 10 per cent added alcohol to straight-run gasoline, producing 'gasohol'. Gasohol has the added environmental advantage of producing a lead-free fuel which also produces less pollution overall (NO_x, CO, etc.).

SUBSTRATES

The major substrates for the production of ethanol by biological means are sugars, starches and cellulose.

Sugars
Ethanol is readily produced by fermentation using yeasts with a sugar substrate. The sugar substrate can be obtained from a number of sources:

- *sucrose*, directly obtained from cane or beet
- *molasses* (sugar waste) produced during the manufacture of cane or beet sugar for use as a sweetener for human consumption
- *glucose* produced from a starch source, for example from cereals like maize, wheat, sorghum, and barley, or from root crops like potato and cassava

- *glucose* produced from cellulose or cellulosic wastes
- *sugars* produced from other wastes like sulphite liquor from papermaking.

The production of potable ethanol for human consumption on a bulk scale (for use in beverages like gin and vodka) must be carefully controlled at all stages, but in the production of ethanol as a substitute liquid fuel, traces of fermentation by-products which would be unacceptable for human consumption may be tolerated.

The largest programme for the production of fuel ethanol directly from sucrose by fermentation is the Brazilian alcohol programme. Sugar cane grows readily in tropical and semi-tropical climates if there is an abundant water supply, and can be planted to allow mechanized harvesting. Fairly simple pressing of the cane produces a sucrose syrup which can be used directly on site after standardizing sucrose concentration as a substrate for *Saccharomyces cerevisiae*.

One major agricultural advantage of sugar cane is that the crop does not have to be dug up and replanted, since it is a semi-perennial, self-regenerating crop, and there are examples of cane stands still producing well up to 20 years after planting.

In temperate climates (Europe, N. America) sugar cane is not a viable crop, and sucrose is produced for culinary purposes (sweetening) using beet. Beet has to be dug up to harvest, it requires re-planting each season, and the extraction of sucrose from it is a complicated process. The arable land area in Europe, for example, is not sufficient to produce enough sugar both to satisfy the demand as a sweetener and to allow major fuel alcohol production.

With the yeast/sugar fermentation process, provided that the *Saccharomyces* strain is carefully selected, strict sterility is not a restricting factor for the production of fuel alcohol, simple boiling of the substrate is sufficient. In addition, the yeast produced during one fermentation can be separated from the broth and re-used in the next batch, trace by-products due to yeast mutations not necessarily being detrimental to the ethanol end use.

Starches
In North America, starch from maize is used to produce fuel alcohol. The corn starch is enzymically converted to fermentable sugars using amylases and amyloglucosidase. Alternatively a chemical conversion can be used to produce a glucose substrate.

The Brazilian government has started a major investigation of a variety of starch-containing crops which may be suitable for fuel alcohol production, and along with cassava and sweet sorghum, various nuts (palm, cocoa, etc.) are being investigated.

Cellulose
Cellulose from wood, straw, and other cellulosic wastes (bran, etc.) can be enzymically converted to glucose for use as a substrate for *Saccharomyces* sp. The problem with wood is that the cellulose is closely associated with hemi-cellulose which must either be broken down or removed to allow the cellulases access to the cellulose substrate. Current processes involve milling the wood to a small size (2–3 mm) followed by acid treatment to break down the hemi-cellulose. The

lignin must also be removed before hydrolysis, and the treated cellulose can then be converted to glucose.

Hemi-cellulose is a major drawback in attempting to use wood as a cellulose source; the energy costs required in physically breaking the material into a small size makes the use of wood as a glucose source only marginally economic.

Acid hydrolysis can also be used to produce glucose, but the conversion is lower than the enzymic process, and is not economically viable as a source of fermentable sugars.

FERMENTATION PROCESSES

Batch Processes

The yeast/sugar fermentation has been extensively developed for the production of beer and wine, and for the batch production of fuel alcohol a very similar process is used.

Anaerobic fermentation of yeast/sugar produces alcohol and carbon dioxide, but aerobic fermentation only produces yeast cells with little alcohol. However, in the anaerobic process, some dissolved oxygen is necessary to provide for cell replenishment, and a dissolved oxygen concentration of about 10 per cent of saturation is optimal for maximum conversion of the substrate to alcohol.

The substrate concentration is typically 12–16 per cent sugars, together with essential inorganic additives and trace elements (nitrogen, phosphate, Mg, Na, K). Inoculation of the boiled, pasteurized (not severely sterilized) substrate with 10–15 per cent by volume of yeast produces a maximum ethanol concentration of 8–12 g/L after 14–20 h. Since the fermentation takes place at pH 3–5, open tanks are acceptable.

The major problem with the process is that the alcohol production capacity of the yeast is dependent on two inhibiting factors:

- *substrate inhibition:* high sugar concentrations (above 12 per cent w/w) inhibit the process, and alcohol productivity in the early stages of batch fermentation is low until the sugar concentration falls below about 10 per cent
- *product inhibition:* high alcohol concentrations (above 12 per cent) inhibit further production, and alcohol productivity also falls in the later stages of the batch process.

Modified strains of *Saccharomyces* have been developed in the laboratory which can tolerate alcohol concentrations up to 16 per cent.

Continuous Processes

Productivity levels using the continuous stirred tank fermentor (CSTF) for alcohol production are much higher than in the batch process (6–8 g/L/h continuous; 1–4 g/L/h batch). The size of equipment required is also reduced, and labour requirements are considerably reduced. However, the CSTF is, of necessity, a more complicated piece of equipment, since more attention must be paid to ensuring that no contamination of the system takes place over long periods of operation. Control equipment is more sophisticated, and a low aeration rate is required to ensure cell renewal (0.1 (v/v)/h), all adding to the capital costs.

Maximum productivity occurs at sugar concentrations of 8–10 g/L with a cell concentration of 11 g/L. Alcohol concentration in the exit stream is of the order of 10–14 per cent (v/v).

An increase in productivity can be achieved by using a cell recycle CSTF. The yeast cells are separated from the exit stream and recycled back to the fermentor. In this operation, cell densities of 50–70 g/L are maintained, and productivity increases to 11–20 g/L/h alcohol. Exit stream concentrations are again of the order of 10–16 per cent (v/v).

DOWNSTREAM PROCESSING

The recovery of alcohol from the fermentor broth is carried out using fractional distillation techniques. A typical recovery scheme is shown in Fig. 9.7.

Cell Separation
For the production of fuel alcohol it is normal practice to recycle the yeast cells, either as an inoculum for the next batch or to maintain a high cell population in a CSTF system. Normal practice is to use a disc type of centrifuge, similar to that shown in Fig. 6.9, which can be operated to give a semi-continuous recycle of cells and a continuous stream of alcohol solution.

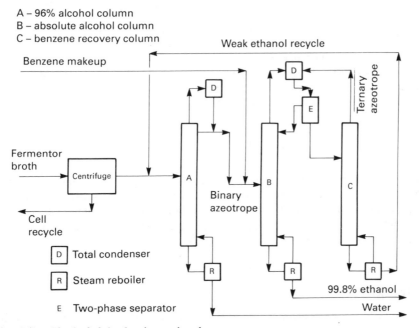

A – 96% alcohol column
B – absolute alcohol column
C – benzene recovery column

Fig. 9.7 Alcohol dehydration using benzene.

Distillation Train
Ethanol for use as a chemicals intermediate is required in a highly pure form (99.8 per cent w/w). For use in the internal combustion engine a concentration of 99.2 per cent (w/w) ethanol is satisfactory.

Ethanol and water form a minimum-boiling-point mixture (*binary azeotrope*) at 95.7 per cent (w/w) ethanol, and since for this mixture the vapour composition in equilibrium with the liquid is the same (95.7 per cent), further concentration is impossible using a single atmospheric pressure distillation column.

The traditional process for the production of anhydrous ethanol uses a third component, usually benzene, which forms a minimum-boiling *ternary azeotrope* of 74.1 per cent benzene: 18.5 per cent ethanol: 7.4 per cent water. Figure 9.7 shows a traditional distillation train for the production of anhydrous ethanol using benzene as the third component.

The dilute alcohol stream from the fermentation (10–12 per cent) is fed to the 96 per cent alcohol column (A) where the *binary azeotrope* is taken off the top and virtually pure water from the base.

This binary azeotrope is fed to the absolute alcohol column (B), with the addition of makeup benzene, where the *ternary azeotrope* is taken off the top and the vapour totally condensed and cooled to about 25 °C. A dilute alcohol mixture is taken from the base of this column and recycled to the 96 per cent column (A) with the fresh feed.

In the liquid state, the *ternary azeotrope* forms two phases, and the lighter, benzene-rich layer (about 65 per cent at 25 °C) is returned as reflux to the column. The heavier layer is fed to the benzene recovery column (C) where the *ternary azeotrope* from the top is recycled to column B (the bottom product to A). The alcohol and benzene losses are minimal and operation takes place at atmospheric pressure.

In order to reduce operating costs, a two-column system can be used with n-pentane as the third component.[8] This system, however, operates under pressure (about 3 atmospheres) and requires better control systems, but the overall energy requirements in terms of steam use are lower.

Single-cell Protein (SCP)[12–15]

The term single-cell protein refers to the biomass derived from a variety of substrates with a view to producing material which can be used as a protein source for human or animal nutrition. Whatever organism is used must not only compete commercially with existing foods or food supplements, but must also meet the relevant food safety and hygiene standards.

SUBSTRATES

The major substrates which have been investigated for SCP production are hydrocarbons, alcohols, and carbohydrates, as well as a variety of waste material from the chemical and food industries.

Spent sulphite liquor from the papermaking industry has been extensively investigated and achieved moderate success using organisms such as *Candida utilis* and fungal organisms specifically aimed at the cattle feed market.

Waste from food processing operations (whey, potato processing waste) has the disadvantage of variation in seasonal supply, and although a number of processes have been developed with these substrates using *Saccharomyces*, *Candida*, and *Kluyveromyces* strains, commercial viability has suffered due to uncertainty of substrate supply. The use of corn steep and other waste liquors from maize processing unfortunately has to compete with the production of antibiotics (penicillin in particular), and since the market value of antibiotics far exceeds that of cattle feed, these substrates must compete with a potentially higher value end-product if SCP from such sources is to be economically viable.

Hydrocarbons and other petrochemical substrates have the advantage of consistency of supply and of relatively steady price. Processes using n-paraffins as substrates for yeasts were developed in the 1970s in Europe and Japan[12, 13] but failed to reach commercial fruition due to consumer fears regarding possible carcinogenic residues which were difficult to remove during the purification of the harvested cells.

Purification of the cell mass is likely to be made easier if a gaseous feed is used (methane, ethane, ethylene), and a number of processes have been investigated using gaseous substrates. However, due to the low solubility, high oxygen demand, and explosion and fire hazard, much of the more recent work on petrochemical sources has concentrated on the use of alcohols, particularly methanol and ethanol.

Most of the above mentioned developments were specifically aimed at the production of SCP for use in animal nutrition. Rank Hovis McDougall developed a process for the growth of *Fusarium graminearium* using food grade glucose[15] aimed at the production of the mycelium for human consumption. The aim was to reproduce a product for use as a meat substitute in terms of texture and appearance.

ICI 'PRUTEEN' PROCESS

Developed by the Agricultural Division of Imperial Chemical Industries, this process uses methanol as a substrate for the growth of the *Methylophilus methylotrophus* bacteria to produce a protein powder specifically for animal feeding.

The plant installed at Billingham (N. England) is an aerobic, aseptic process with a capacity of 50 000 t/yr of SCP. The fermentor has a volumetric capacity of 1500 m^3, making it the largest aseptic fermentor in current operation. The process is operated continuously and sterility must be strictly maintained, otherwise the product may become unfit for animal feeding. Contamination also reduces the yield of *Methylophilus methylotrophus* which adversely affects the economics of the process.

A general process diagram is shown in Fig. 9.8.

Substrate
The major substrate is methanol, supplemented with phosphoric acid, calcium, magnesium, iron, and trace elements. Nitrogen is added as ammonia gas with the

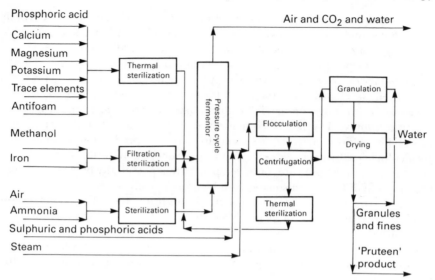

Fig. 9.8   ICI 'Pruteen' process.

air stream. In order to minimize the raw materials usage, the fermentor broth is re-sterilized and recycled back to the fermentor after separation of the cells.

Sterilization
Initial sterilization of the fermentor and associated pipework and fittings prior to startup is carried out using steam under pressure. Because of the need for thorough sterilization prior to continuous operation, the process is carried out under computer control to minimize any possibility of sterilizing sequence errors.
  The liquid feeds to the process are sterilized using two different methods:

- the liquid stream containing the supplements (phosphoric acid, etc.) and the recycled broth liquor are thermally sterilized using conventional techniques
- the methanol stream is sterilized using biological filters (membranes). The reason for this is that the methanol is co-produced on-site using petrochemical process techniques, and does not have a high bacterial count.

  The air (together with the ammonia) is sterilized using conventional fibre filter techniques.

Fermentation
Because of the size of plant required for economic operation (25–150 000 t/yr) the fermentor volume required is about 1500–2000 m^3. This volume could have been provided by using a number of conventional CSTFs working in parallel, but capital costs would then have been enormous. Also, because of the need to ensure a high level of sterility for long, continuous operating periods, it was desirable to have only one fermentor which could be closely controlled. This requirement for a

single fermentor led to the development of the *pressure cycle fermentor*. This consists basically of an air-lift fermentor (see Fig. 4.3), but with a capacity 10 times the largest operating penicillin fermentor and about 30 times the largest air-lift fermentor, the design is somewhat special. An outline diagram of the pressure cycle fermentor is shown in Fig. 9.9.

The plant installed at Billingham is 7.0 m in diameter and approximately 50 m high. The pressure at the base due to the static head of liquid is about 5 atmospheres, and with overpressure at the head of 1–1.5 atmospheres, the base pressure is 6–7 atmospheres.

This has the advantage of allowing a high dissolved oxygen concentration at the base, and the redistribution devices within the 'draft' tube ensure good air redistribution and substrate mixing together with carbon dioxide disengagement.

The fermentor body is welded in one piece throughout, eliminating possible sterility leaks at flanges, and since there is no agitator shaft seal, possible contamination from this source is also eliminated. The heat of fermentation is removed by heat transfer devices placed in the base of the liquid recycle downcomer tube.

Operating variables are shown in Table 9.4.

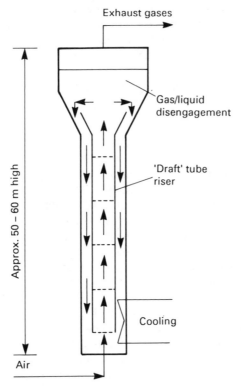

Fig. 9.9 Pressure cycle fermentor.

Table 9.4 Operating conditions for a pressure cycle fermentor

Temperature	37 C
pH	6.7
Methanol concentration	10 ppm in broth
	< 1 ppm in exit
Air requirement	13 000 m^3/t 'Pruteen'
	0.68–0.9 (v/v)/min
Dilution rate	0.1–0.2 h^{-1}

Broth Separation

The concentration of cells in the fermentor broth is about 3 per cent (w/w), and to aid separation the exit stream is treated with phosphoric and sulphuric acids and steam. This causes the cells to flocculate, and separation is carried out using a continuous centrifuge.

The 20 per cent (w/w) cell concentrate is fed to the drying system, and the separated liquor is re-sterilized and recycled back to the fermentor.

Drying

The 20 per cent cell concentrate is neutralized with sodium or calcium hydroxide. The drying system uses a flash process with hot furnace gases and the concentrate is mixed with recycled (dry) granules and fines in a granulator. The wet granules are introduced into the hot furnace gas stream and collected in a cyclone separator.

The fines and some of the granules are recycled back to the granulator, the excess being cooled and stored as product.

Conclusion

The processes described in this chapter have covered only a limited selection of the many manufacturing processes in existence. The penicillin process is a good example of a process which has been highly developed in both the fermentation stage and in the downstream recovery operations. However, the process was developed in the late 1940s before the availability of strong reverse osmosis membranes, and if the process were being developed today, membrane processes would be being considered as an alternative to the original precipitation and extraction processes. Effluent treatment is an example of a process which is operated almost entirely non-aseptically, and involves the handling of large quantities of liquids and solids. Fuel alcohol production uses conventional technology well known in the alcoholic beverage and petroleum chemicals industries. The ICI 'Pruteen' process required a new approach to fermentor design because of the long-term sterility requirement, but the success of the pressure cycle fermentor has led to a continuing re-assessment of general fermentor design and control.

From these limited examples, I hope that the role and use of the various unit operations in process engineering has become clearer, and that in looking at other processes described in the general literature, the engineering descriptions will be more understandable.

References

1. A.L. Elder (ed.), *The History of Penicillin Production*, Chem. Eng. Prog. Symp. Ser. No. 100, vol. 66, AIChemE, New York (1970).
2. G.J.M. Hersbach, C.P. van der Beek, and P.W.M. van Dijck, 'The penicillins: Properties, biosynthesis and fermentation', in *Biotechnology of Industrial Antibiotics* (ed. E.J. Vandamme), Marcel Dekker, New York (1984).
3. G. Hamer, 'A biotechnological approach to the treatment of wastewater from petrochemicals manufacture', in *Effluent Treatment*, Inst. Chem. Engrs Symp. Scr. No. 77, p. 87, Rugby (1983).
4. D.A. Stafford and S.P. Etheridge, 'The anaerobic digestion of industrial wastes, farm wastes and sewage sludges', in *Effluent Treatment*, Inst. Chem. Engrs Symp. Ser. No. 77, p. 141, Rugby (1983).
5. D.A. Bailey and J.B. Rhoades, 'The water service and its impact on industry', in *Effluent Treatment and Disposal*, Inst. Chem. Engrs Symp. Ser. No. 96, p. 1, Rugby (1986).
6. G.K. Anderson and C.B. Saw, 'Applications of anaerobic biotechnology to waste treatment and energy production', in *Effluent Treatment and Disposal*, Inst. Chem. Engrs Symp. Ser. No. 96, p. 137, Rugby (1986).
7. H. Rothman, R. Greenshields, and F.R. Calle, *The Alcohol Economy*, Frances Pinter (Publishers), London (1982).
8. K. Esser and U. Schmidt, 'Alcohol production by biotechnology', *Proc. Biochem.* **17**(3), 46 (1982).
9. S.K. Tangnu, 'Process development for ethanol production based on enzymic hydrolysis of cellulosic biomass', *Proc. Biochem.* **17**(3), 36 (1982).
10. C. Black, 'Distillation modelling of ethanol recovery and dehydration processes for ethanol and gasohol', *Chem. Eng. Prog.* **76**(9), 78 (1980).
11. N. Kosaric, Z. Durnjac and G.G. Stewart, 'Fuel ethanol from biomass— production, economics and energy', *Adv. Biochem. Eng.* **20**, 119 (1981).
12. B.M. Laine, R.C. Snell, and W.A. Peet, 'Production of single cell protein from n-paraffins', *Chem. Eng. London* No. 310, 440 (1976).
13. A. Einsele, 'Biomass from higher n-alkanes', in *Biotechnology* (eds H.J. Rehm and G. Reed), Vol. 3, *Microbial Products, Biomass and Primary Products* (Vol. ed. H. Dellweg), Verlag Chemie, Weinheim (1983).
14. S.R.L. Smith, 'Single cell protein', *Phil. Trans. Roy. Soc. London* **B290**, 341 (1980).
15. G.L. Solomons, 'Single cell protein', in *General Review in Biotechnology* (eds C. Stewart and I. Russel), CRC Press, Boca Raton (1983).

Index